IN FULL CRY

a

IN FULL CRY

BY

RICHARD MARSH

Author of
"The Beetle: a Mystery," "Curios: the Strange
Adventures of Two Bachelors," "Frivolities,"
"The Crime and the Criminal," "The
Datchet Diamonds," "The House
of Mystery," Etc., Etc.

LONDON
F. V. WHITE & CO.
14 BEDFORD STREET, STRAND, W.C.
1899

CONTENTS

CONTENTS

IN FULL CRY

IN FULL CRY

CHAPTER I

THE PRICE OF BLOOD

"THAT'S a nice little bit of money, that is. I wouldn't mind the handling of it."

"Then you can handle it for me. I wouldn't touch it; no, not if it was two thousand I wouldn't touch it; no, not if it was two 'undred thousand I wouldn't. I ain't no blooming 'angman, and I ain't going to do no blooming 'angman's work. Let them do it as likes it."

The speaker, thrusting his hands into the pockets of his ragged breeches, hunched up his shoulders as if he were cold.

"Yet you could do with a bit, couldn't you, Sam? Two hundred pounds is two hundred pounds—it ain't to be picked off every crossing."

"I ain't saying as 'ow it is, am I? But when it comes to 'anding a chap over to them blarsted slops for them to wring his neck and you gettin' two 'undred for givin' 'im away, why, I says, and I says again, let them do that sort of truck as likes it. I don't. S'posin' it was you as killed the bloke and I was to say to you, 'Nosey, I'm goin' to earn two 'undred quid, and I'm goin' to get yer 'ung to do it,'

A

what would you say to me, eh? What do you
think? What would you think of me—that's what I
arsk?"

"But I didn't kill the bloke, wurs luck!"

"Why wurs luck?"

"'Cause I should have had sumfink if I 'ad—sumfink what might 'ave brought me in a piece or two.
Look 'ere." He pointed with outstretched finger to a
line on the placard, reading the words with difficulty,
one by one. "The cove what done for him did get
sumfink for doin' it. 'An—ex-panding—gold—
bracelet — with— large—ruby—set—in—circle—of—
pearls. Gold — ring — with —large—single—stone—
diamond — mon-o-gram— inside—J.—H.—S.' Them
two corst money. Who's to say they wouldn't bring
you in that couple of 'undred quid?"

"What are yer sayin'? Go on! 'Tisn't likely!
There you are again! Tell yer I wouldn't 'ave 'em,
not if you was to take 'em out of your pocket and
say, 'Sam, 'ere's a free present for yer!' They'll
'ang the chap as they're found on, certain to.
Directly he tries to place 'em he's a goner. Hullo,
Pollie! Here's some spicy readin'. Two 'undred
pound a-goin' a-beggin'. Nosey, 'ere, says 'e'd like to
earn it, but I says, not me."

The words were addressed to a girl who had
approached them from behind, and who joined them
in their examination of the placard which dangled
on a nail against the whitewashed wall. The man
spoken of as Nosey, turning, looked at her with his
bloodshot eyes for a second or two in silence.

"'Ow's things with you? Ain't seen you about
just lately."

"No," said the girl, "you haven't."

Her voice was full and clear, her manner brusque,
as if she wished him to understand that she did not
invite his conversation. Sam, his hands always in

his breeches' pockets, walked towards the fire, speaking as he went.

"Well, there you are, Pollie; there's two 'undred quids a-goin' a-beggin'. P'r'aps you'd like to make it yourn. There's many a gal as 'as sold a chap for less than that."

Nosey spat on the floor.

"Less? Why, there's 'eaps of women as 'as sold their blokes for the price of 'arf a pint! Yes, and for a smell at it—'eaps!"

The two men began to warm themselves before the blazing fire. Others, entering the room, joined them; they were shivering with cold, greeting the dancing flames with hungry, yet timid, eagerness. The girl continued to stare at the placard which was hung against the wall.

It was a police notice. It was headed "Murder! £200 Reward," stating that whereas John Howard Shapcott was murdered in his rooms in Embankment Chambers on the night of Tuesday, December 22, by some person or persons unknown, the sum of £200 would be paid to whoever gave such information as should lead to the capture and conviction of the criminal or criminals. The girl read the first part of the notice two or three times over, as if she was unable to altogether grasp its meaning. She muttered to herself.

"Murdered! Well, what of it? I wish it was me instead of 'im—there's nothing in being murdered. I'd sooner be dead than living like I am, that I would."

She tossed her bare head, with its shock of jet-black hair; drew her scanty shawl tighter round her; glancing at the group about the fire with flashing eyes, like some hunted, startled thing; and shuddered—not with cold, but with some emotion which she found it difficult to master. She continued her examination of the poster; her attitude, which had suggested pre-occupation of mind, suddenly changing to one of unmistakable and lively interest —interest which was so intense as to be painful.

The placard went on to state that certain articles had been stolen on the night of the murder, and pawnbrokers were requested to give immediate notice to the police should either of them be offered in pledge. Two of the articles in question were described in some detail; it was this description which seemed to rouse the girl to such sudden interest.

" 'An expanding gold bracelet with large ruby set in circle of pearls. Gold ring with large single stone diamond; monogram inside, J.H.S.' My God! my God!"

As she read the description she staggered against the wall in a state of partial collapse. In a moment or two, drawing herself a little away, peering at the printed words, she read them over and over again, as if they had for her a fascination from which she could not escape. Seeming to satisfy herself, by means of a furtive glance over her shoulder, that the group about the fire were oblivious of her proceedings in their enjoyment of its warmth, lifting her skirt in front, from some receptacle beneath she took a strip of dirty white rag. She clutched it tightly in her hand, as if she were afraid that someone might discover it in her possession; then quickly untying one of the corners she discovered something within which one would not have expected to find in such a covering. It was a gold bracelet, of lattice work, so flexible that, when closed, as it was then, it would hardly have admitted a lady's little finger. On one side was a magnificent ruby encircled by pearls. The girl did scarcely more than glance at it; as rapidly as she had untied, she tied it up again. Shifting the position of the piece of rag in her hand, she turned her attention to another corner. From it she took a ring, a gold ring set with a single diamond. This she examined closely, particularly the inside of the circle. There, in a monogram, which was less than usually fantastic, were engraved the letters, plainly to be seen—J.H.S. Her hand fell to her side; her jaw dropped open.

"My God! He done it!"

She stood staring at the placard in front of her, vacantly, in a sort of stupor. Then, recollection coming back to her with a sudden rush, with feverish haste she restored the ring to its uncongenial shelter, and then the rag itself to its original hiding-place. She staggered to a stool which stood by the trestle table; her hands lay wide open on her knees, her back was bowed; she sat with unseeing, staring eyes.

"Gentleman! Gentleman! Gentleman! Why did you do it?"

Presently her figure stiffened; her back straightened; her hands closed; the fists became clenched; she looked at the placard with meaning in her eyes.

"£200 reward! It's a deal of money—a deal. More than I'll ever earn, or see the colour of. It's more than he's worth, if that's the sort he is. I hate him! I hate him! I hate him!"

She repeated the assertion three times, with increasing emphasis, as if desirous that the mere fact of the repetition should make her hatred more.

"Killed him, did he? And never breathed a word! And brought the things he'd robbed him of straight home to me? And him a gentleman! If they'd got me into trouble he wouldn't have cared. They might have done. How did he know? Perhaps that's what he brought 'em for. If I thought—"

Her jaw shut tight; her lips parted, showing the regular white teeth beneath.

"If I thought—"

She drew a long breath, presently announcing that she had put the seal to a resolution at which she had inwardly arrived.

"That I would! Why shouldn't I? He's used me bad, like as I was the muck beneath his feet. And I've done my best for him; yes, more than any other girl'd have done. And what's he done for me?" She shuddered. "Curse him! Told me to go to the devil, did he? Very well; I'll pay him off. I'll go. Says he's sick of me. I'll make him sicker. Wants to know if a girl of my sort ought to have any truck with a gentleman like him, after all I've done for him; I'll give him his answer. Curse him!"

Her fingers kept unclosing and closing. A sort of twitching seemed to attack the muscles of her hands and arms.

"£200! It's a fortune! I'd be a lady with all that money; yes, that I would. I'd have some fun with it, my! And me with hardly a bit of boot to my feet, and not a hat to my head!" Stretching out her feet, she looked savagely at the relics of the shoemaker's art which certainly did not cover them. "And not a crust of bread to put into my belly, nor the money to buy one. I've spent my last brown on getting a doss in here. I couldn't sleep rough through a night like this; by the morning I'd have been froze. I'm pretty near froze as it is, and I'm clemmed with hunger. Clemmed, I am!"

She looked about her with wolfish glances, sufficient witnesses to the reality of the appetite of which she spoke. With the fingers of her left hand she felt for the strip of rag which was concealed beneath.

"I thought there was something up when he give them to me, especially when he only laughed when I asked him where he got 'em from. I knew they was worth money, and it wasn't likely he'd give 'em to me unless he was up to some little game of his own, and him dead stony. If I'd tried to raise a dollar on 'em I'd have got myself in trouble; they'd have said I'd a hand in what he'd been doing, and then, how should I have looked, I'd like to know? I wonder if he give 'em to me so as to get himself well shut of me?'

As she asked herself the question, a great sob, rising in her throat, seemed to choke her. Turning, laying her arms on the table, she pillowed her face on them. Although she uttered no audible sound, her whole form shook with the violence of her emotion.

A young man, coming into the room, paused in the doorway, glancing round to see who the other occupants of the apartment might chance to be. He was about twenty years of age, tall and thin, with closely-cut fair hair. He wore an old cloth cap at the back of his head. A check scarf was wound round his neck. His hands were thrust into the pockets of a pea-jacket which had seen much service. In spite of the fact that his thin face bore the expression of preternatural shrewdness which is the inevitable hall-mark of the London lad whose schoolroom has been the streets, there was about him something which was not unattractive. There was honesty in the frank look of the grey eyes, an odd suggestion of tenderness in the curve of the strong mouth, a winsome independence in the carriage of the well-shaped head.

At sight of the head recumbent on the table he quickly crossed the room. A look of trouble came on his face which was almost comic in its suddenness. Sitting on a stool at the girl's side, he laid his hand upon her shoulder.

"Pollie, what's up?"

Raising her face, she looked at him with a glare which was reminiscent of a wild animal.

"Everything's up."

"Where's the Gentleman?"

"I don't know where he is; I hope he's dead."

The cloud on the lad's face grew blacker. His upper lip trembled. He was still for a second or two.

"What's he been doing now?"

"Doing?" she laughed, not pleasantly. "He's sent me packing, that's what he's been doing. He's told me to sling my hook. He chucked me out into the street two days ago, he did, and said he never wanted to see my face again. But I'll pay him; when he does see it again it'll be the sorriest day in all his life."

"The brute!"

The girl resented the epithet with sudden unreasoning rage.

"Don't you go calling him no names—I can do all the calling of names that's wanted, so you can take the tip from me. Don't I say I'll pay him? And so I will. For good and all!"

The lad leaned closer towards her. He lowered his voice.

"I'll tell you how to pay him best."

"I don't want no telling. Thank you all the same, my lad. Think I'm a fool. When I want to get my knife into a bloke I can stick it in myself—as well, perhaps, as here and there another."

"You can pay him best by marrying me."

She stared as if taken by surprise. Then she laughed again—again not pleasantly.

"If it's all the same to you, I don't happen to be takin' any—not to-day, I don't."

The lad spoke to the girl with a degree of earnestness which lent to his words a halo of romance which intrinsically they lacked.

"Don't say that, Pollie. You know I've been

saving up this ever so long, and lately I ain't been doing at all badly. I've got twenty-two poun' ten— it's in my pocket at this moment."

The magnitude of the sum amazed her.

"That's a heap of money, Bob."

"We can do with it, you and me together, Pollie, if you'll just say the word. I'll get you a home— a home of your own; straight, I will. I've had my eye on one down in the shop in Walham Green—'a home complete for five pounds,'—that's what it says. Bed, and carpet, and arm-chair, and pots and kettles and all—the thing. And I'll buy a moke and a new barrer—I know where to lay my hand on both the two of 'em, cheap!—and I'll set up a high-class stock, and we'll make our fortune—you can take it from me, Pollie, that you sha'n't want for nothing if you just say the word."

"I tell you I'm not taking any."

"Don't you keep on saying that; don't you, Pollie. I know I ain't a gentleman—"

"No, you ain't."

"But I ain't such a bad sort in my way."

"I never said you was, that I know of."

"And I—I'm fond of you; I love you, Pollie— straight, I do! There's nothing I wouldn't do for you—nothing in the whole wide world."

"When I want anything I'll let you know."

The tears stood in the lad's eyes; he was so much in earnest.

"You think it over before you make up your mind, there's no hurry. Whenever you're ready, I am; I'll keep on waiting."

"It's no good your keeping on waiting, I never shall be ready."

"Pollie!"

"I tell you I sha'n't. I've had one bloke, and I'll never have another; if I die to-morrow or if I live a hundred years. You know me, and you know that what I say I mean."

"But he chucked you out into the street."

"Yes, and I'll pay him for it. Don't I tell you I'll pay him for it—once, and once for all." Her eyes wandered towards the placard which hung against the wall; her fingers gripped something which was concealed beneath her skirt, "but that'll make no difference, so far as you go. I've had my share of men; they're off. I tell you I don't want to hear no more about it."

"But, Pollie—"

"Shut it! If you keep on there'll be trouble. You've asked me a straight question, and I've given you a straight answer, and that's enough."

The lad looked as if he himself was not so sure. But apparently what he saw in her face caused him to conclude that, on this occasion, discretion might be the better part of valour. He gulped down something in his throat, with an odd air of swallowing something disagreeable. Rising, turning his back on her, he began to study the placard which hung against the wall, though, possibly, without realising what was in front of him.

If that was so, he was roused to the reality of the case by a voice which proceeded from the group which still hung round the fire.

"Tidy lump of oof, two hundred quids, eh, Bob? About suit you."

"It would that."

"Would you like to earn it? You've only to get the chap's neck stretched, and it's yours."

"If I knew who it was I'd earn it fast enough. A chap what had done a murder I'd give away, whether there was two hundred pound to come from it, or whether there was nothing."

"Anyone would think you'd done it yourself to hear you talk."

The lad wheeled round with a burst of sudden passion.

"What do you mean by that? You say again I done it."

The man by the fire regarded him as if astonished.

"I ain't said once you done it, not yet I haven't. What's the matter with the kid?"

"Then don't you hint at nothing; I won't have it!" The lad brought down his fist upon the trestle table with a bang. "The chap what does a murder ought to be hanged, I don't care who he is; and if there's nobody else would hang him, why then I would, so now you know."

Something in his words or tone seemed to touch the girl on a sensitive point. Springing to her feet, she confronted him with a show of anger which was equal to his own.

"A lot you'd do! You're the sort to talk, you are. Why, I tell you there's heaps of blokes as want murdering, heaps and heaps of them, and some as I'd only like to get the chance to murder, and perhaps you're one." There was a laugh at this. "Howsomever that may be, you may take this as gospel truth, that there are chaps as does murders whose shoes you ain't fit to wipe; no, nor never will be!"

In the silence which followed, one of the men who had been standing by the fire went out of the room. He was heard addressing someone without.

"Hollo, Gentleman! Here's Pollie Hills inside, waiting for you."

A loud, ringing voice replied, the voice of an educated man. Each word he uttered was clearly heard within, as probably it was meant to be.

"Pollie Hills be—sanctified! Is that unfortunate creature to dog my steps wherever I go! Too bad! This comes of a gentleman condescending to a person of the sort. Take my advice, my good fellow: never

have an affair of the heart with a female of a lower
rank than your own—if it is possible to imagine such
an animal. I'm afraid that either I must give place
to Miss Hills, or Miss Hills must give place to me;
and the latter for choice."

The men about the fire were staring at the girl with
grinning faces. She had turned a vivid scarlet. Her
eyes flamed. Without a word, she rushed from the
room, taking, even as she tore along, something from
its hiding-place beneath her skirt, something which
was in a dirty white rag.

CHAPTER II

THERE entered, almost in the same instant as the girl vanished, so that she must have brushed against him as she passed, a man of a type not frequently seen in that stew of incongruous elements, a London "doss-house." He was very tall, and he held himself very straight. The absence from his bearing of that nearly universal feature of the carriage of the patrons of such establishments, the boneless back and the invertebrate neck, would in itself have marked him out as singular. His face, if not exactly handsome, was at least remarkable: it was the face of a man who had escaped being a genius by a hair's-breadth only. The lofty brow and large, well-shaped nose betokened intellect; the loose lips were sensual; while in the small, bright, continually-restless eyes was the very spirit of whimsicality. He kept his cheeks and chin and lips clean shaven, a habit which, as much as anything else, had gained for him his sobriquet of "Gentleman." His hair, tinged with grey, and parted scrupulously on one side, he wore a little long. The chief articles of his attire were a top hat, which had had a long and most adventurous career, and a capacious frock coat, which had seen even more service than the hat. This latter garment was buttoned to the throat. A narrow black ribbon, protruding from beneath a ragged collar, strayed loosely over the front of the coat. To complete his costume,

13

he carried on an ordinary piece of grocer's twine an
eye-glass, chipped and scratched, with which, when
it was not fixed in his eye, his fingers were continually trifling. Despite the ostentatious evidences of
his dire poverty, and of the many buffetings which
the world had given him, there was about his looks
and bearing, his voice and manner, an airy ease, an
insolence, an assurance, an importance, a self-con-
fidence, a suggestion of superiority which one is wont,
as a rule, to associate with success rather than failure.
Even as, fitting his glass in its place, he entered with
just a touch of swagger, there was about him a
flavour of condescension which did not become him
ill. Removing his hat, he treated the occupants of
the room to a sweeping bow, standing stock still as
he did so, to give the action more effect.

"Gentlemen, how wags the world? At least, in
such weather, better a fourpenny doss than a
gratuitous arch."

The man spoken of as Sam made room for him
beside the blaze.

"That's right, Gentleman, you come and 'ave a
warm. A fire's a fire to-night, it is."

The Gentleman, moving forward, was about to take
advantage of the offer, when the lad Bob, thrusting
himself forward, confronted him with clenched fists
and lowering brows.

"You —— "

The newcomer, showing no surprise at being greeted
in such a fashion, merely regarded him sideways with
twinkling eyes.

"My good sir!"

"Don't try to come that with me; I won't have it.
You're a —— ——! and a —— —— ——! and now
you've got it! Hear?"

"I hear very well. That is a faculty which still
survives—alas!"

"And you're a —— ——! and —— ——! and
I wouldn't use you to wipe the gutter with!
See?"

"My frankly-spoken young friend, I do see—
clearly. And what is it I am meant to do? What
position ought I to take? Am I supposed to knock
you down, or to demand from you the satisfaction of
a gentleman? Because, in either case, I am afraid
I shall have to be to you the occasion of disappointment."

"I don't want none of your tall talk, nor none of
your soft sawder neither; I don't understand it, and
I won't have it; but just you take this from me: if
you don't care what you're doing to Pollie Hills, I'll
break every bone in your —— body, and knock
every tooth down your —— throat! and don't you
forget it."

"Now I gather the intention. So this is merely an
illustration of chivalry as a modern survival. The
modernity is unmistakable. I assure you I'm not
likely to forget, thanks.—What have we here?"

His eye had been caught by the police notice hung
against the wall. He crossed to examine it more
closely. Sam's voice followed him.

"That's something in your line, Gentleman.
There's two hundred quids offered for the chap as
done that bit of work. We was just a-argyin' as to
whether a bloke ought to give away the chap as
done it so as to earn them quids. Now, what do
you say, as a gentleman?"

The Gentleman was standing before the notice, his
legs apart, his hands behind his back, his head a
little on one side, his expression growing more and
more whimsical the clearer he grasped the purport
of the placard.

"Murder! Dear, dear! how sad! John Howard
Shapcott! I seem to remember the name! A fellow

of infinite jest, infinite! It doesn't say how he was murdered, this unfortunate Mr Shapcott."

"Shot, he was. I see it in the paper."

"I know; I fancy I've seen it in the paper also. I believe there are cases in which it is perfectly painless to be shot, which is a blessed dispensation of Providence. Something taken, too—robbery and murder. Deplorable, quite!"

"And there's two 'undred quids offered for the bloke as done it. Now would you say as a chap did ought to earn the money, seein' as 'ow 'e 'ad the charnce?"

"Certainly; why not?"

"What! 'and 'im over to them there blarsted slops, for them to twist 'is neck?"

"He's a criminal, my friend, a criminal. Don't allow that point to escape your observation. It's indispensable, for the safety of society, that every man's hand should be against a criminal."

"I don't see it myself. It's blood money, that's what it is, and I shouldn't care to 'ave none of it come my way."

"You're singular, my friend, you're singular. The world is not of your opinion, or the risk of our being murdered in our innocent beds would be greater than it already is. Now I've invested a portion of my modest fortune in the purchase of a newspaper which contains some most interesting particulars of this truly deplorable occurrence."

Seating himself on the edge of the trestle table, resting one foot upon a stool, he took a paper from his pocket which he slowly unfolded.

"Here it is: 'The Tragedy in Embankment Chambers.' Wonderful intelligence is displayed in the conduct of the modern press. Here's a man, a journalist, who seems to know so much about the case, and what I would call its ramifications, that one is almost inclined to suspect him of having had a hand in the affair himself, which would, of course, be absurd. The police, however, do not seem to possess a clue to the miscreant who did the deed. But they will have, they will have. Someone will claim that two hundred pounds—sure to. Murder will out, it's proverbial. And murderers are sure to get hung, which great and incontrovertible truth is, perhaps, not so commonly known. It seems that Mr John Howard Shapcott was a most estimable character, as men are generally shown to be, most clearly, after they are dead. Without an enemy in the world — which under similar circumstances, again, is usual. Our enemies die with us, but are instantly resurrected in the form of friends. He has left a fortune, a large fortune, the newspaper says, an accident which is not so commonplace. Now, if by some quaint chance he should happen to have left any portion of his fortune to the man who murdered him, I should find that very funny, wouldn't you? I suppose, gentlemen, none of you have committed a murder yet, as an incident in your career?"

His audience hardly seemed to know how to take his question, whether to be offended or not. Presently a man with red hair and beard, and a broken clay pipe held in the corner of his mouth, took upon himself to answer the inquiry so far as he was personally concerned.

"I ain't, guv'nor; I ain't got as far as that, not yet. But there's no knowing 'ow far you may get afore you're done. My Gawd, no!"

"That's true; one never does know how far one may get before one's done. I wonder what it feels like to kill a man. What it feels like, for instance, to see a man lying dead at your feet, and to know that you laid him there."

B

"Ordinary. I've killed a pig. I'd make no more of killin' a man, if I'd got to."

"You've killed a pig? I, also, have killed a pig."

"You! You ain't no butcher, guv'nor."

"Who knows?" The Gentleman shrugged his shoulders, regarding the speaker with his whimsical smile. "We are all of us something which no one suspects. Why shouldn't I be a butcher?"

Before anyone replied, the deputy—the "doss-house" manager—came into the room.

"There's someone wants a party of the name of Polhurston; anyone here of the name of Polhurston?"

The Gentleman's back was towards the door. As the name was uttered, the fashion of his countenance was changed. He started. The newspaper he was holding fluttered. Then he sat motionless, gripping the paper with both his hands.

The deputy's inquiry went unanswered.

"Do you hear, you chaps? Anyone here of the name of Polhurston? Blaise Polhurston, ain't it?"

Again the man upon the table started.

"Precisely. Is Mr Blaise Polhurston here?"

The reply, and further question, came from an individual who had followed hard upon the deputy's heels; a man between fifty and sixty years of age, who carried a pair of *pince-nez* in his hand, which, as he spoke, he lifted to his eyes.

As he spoke the Gentleman drew a great breath. He bit his lip, then, seeming to arrive at a sudden resolution, stood up and turned.

"I am Blaise Polhurston."

The new-comer regarded him through his glasses.

"Mr Blaise Polhurston of Trevennack?"

"Precisely. Glad to see you, Baynes, after all these years. Don't you know me? Am I changed so wholly out of knowledge?"

"On the contrary, I should have known you anywhere." The speaker advanced with outstretched
hand. "The pleasure of meeting you, Mr Blaise,
is wholly on my side. It is not my fault it has been
postponed so long. We have advertised for you
everywhere."

The Gentleman shrugged his shoulders.

"Advertised?"

"We have searched for you by every means we
could contrive, but it's only within the last hour
that I have learned where you might possibly be
found. You see, I've come at once. Is there anywhere where I can speak to you in private? Or
will you come with me?"

"I am obliged. This is my home for to-night.
I am not disposed to turn out again until, in the
morning, in the ordinary course, I'm turned."

"I have something to say to you which should be
said at once."

"Then say it."

Mr Baynes looked round the room. His tone was
significant.

"Here?"

"Why not? And at the top of your voice. It's
a custom of the house. We all discuss our most
secret matters with the full force of our lungs.
We're a lot of poor devils who are waiting for the
ferry to convey us from this kingdom of the damned
to another, which hardly can be worse, and will, at
any rate, be final. What should we want with
privacies?"

"I am sorry to hear you talk like that."

"You needn't. There's the door! You can hardly
expect me to go out of it because you have chosen
to come in."

Mr Baynes' tone was not only dry, it was a little
rueful.

"As I have already remarked, I should have known you anywhere."

The Gentleman, stretching out his hand, touched him on the shoulder.

"Baynes, you score. You're right. Inside and outside I'm the same, yesterday, to-day, and for ever."

"One would hardly have expected a man who has seen so much to have changed so little."

"What we come into the world, that we are, and remain, and go out of it. It is only externally we change; within, we are the same."

"I think you are mistaken. However, that is by the way. Will you come and see me in the morning?"

"No, Baynes, I promise you I won't."

"At least, come out of this room."

"Not I."

"Then step with me aside."

"I'll not stir from where I stand. What you have to say, say out. There's nothing you can say to me I'm not content that all the world should listen to."

Mr Baynes advanced a step nearer. He lowered his voice.

"I've something to say to you—"

"Speak up, man! It's rude to whisper in company. Where's your manners?"

Mr Baynes fell back the step he had advanced. He spoke curtly, with a trace of sternness.

"You have been left a fortune, Mr Blaise."

The muscles of the Gentleman's face twitched. His eye-glass fell from its place—swinging on its piece of grocer's twine.

"A fortune! Me!"

"You!"

"You jest!"

"It is not a subject on which I should be disposed

to jest; nor should I be likely to seek you out merely to shoot at you so singular a shaft of humour. It comes from one from whom you might hardly have expected it."

"My faith, if a fortune comes to me it most certainly will be from a quarter from which I had no expectations. Who's the curious creature?"

"Howard Shapcott."

The Gentleman staggered back against the table. His face took on an extra shade of pallor.

"What? Who?"

"Howard Shapcott."

"You—you jest again."

"I am entirely serious."

"But—but—he's murdered."

"Unfortunately. But that fact will in no way affect the validity of his testamentary dispositions."

"But when I saw him last he refused to assist me even with a shilling."

"When did you see him last?"

"When? Oh, some years ago."

"Probably, in the interval, his mental attitude towards you altered for the better."

"There wasn't time."

"Not in years? The facts show it to have been otherwise. The only will which has been found names you as sole residuary legatee."

"Sole residuary legatee! Shapcott's! It's a nightmare!"

"On the contrary, I confidently believe that it is for you the dawn of a brighter day, and of a happier future."

"A happier future? For me! Baynes, you keep on jesting."

He had sunk upon a stool, his back against the table. His chin had fallen forward on to his chest. He looked the picture of dejection. The good tidings seemed to have affected him more grievously than all Fortune's unkind buffets. Mr Baynes went and laid his hand upon his shoulder.

"Come, Mr Blaise, play the man. You've been, at least, as much sinned against as sinning. I know that as well as you. There's time to change, and here's the opportunity."

"Shapcott's money?"

"Yes, Shapcott's money. There's a good deal of it; if you use it to anything like advantage, it should serve to put black care behind you for ever and a day."

"My God!"

"Money's not everything, as you and I both know; but it's something. Now, let's be off."

"No, not now."

"Yes, now."

"I say no. I'll come to you in the morning."

"Why not come with me at once?"

"Because I'll not. I want to digest your news in my own fashion. In the morning I will come."

"You promise?"

"When I say that I will do a thing, the saying does not need the support of any promise, Mr Baynes. Some years ago I said I would do something. I have done it, as you are aware, and yet I never promised." He stood up. Some of his former manner returned to him. "Now, let me understand what my position really is. I'm a man of means?"

"You are."

"Of fortune?"

"Yes, of fortune."

"And yet, if you will forgive my entering into minute details, do you know that, at this moment, I have not as much as sixpence in my pockets? Not that it would be of much use if I had, for my pockets are but holes. I carry my money, when I have any, in my hat. Permit me, Baynes, to show you."

Removing his hat, he took from the greasy lining a piece of dirty rag, the double of the one which was the property of Pollie Hills.

"When one lives in curious ways, Baynes, one falls into curious habits." He unfolded the rag. "You perceive, two pence! the balance of a silver sixpence which an old lady gave me for fetching her a cab. Fourpence furnished me with a home to-night; I was reserving the twopence to provide myself with supper. Now, do you think, Baynes, that, by way of inspiring me with confidence, and proving that you are not jesting, it would be possible for you to transform that twopence into a sum possessed of a higher purchasing power?"

"Mr Blaise, why won't you come with me at once?"

"Is that your answer?"

"Would a five-pound note be of any use?"

"Not the slightest. No one has so much specie hereabouts; and, if they had, they would not change a five-pound note for me."

Mr Baynes took out a sovereign purse.

"Here's four pounds."

"That's better, Baynes." The coins were handed over. "Since I've held a sovereign in my hand the good God alone knows how many days it is. Now, Baynes, good-bye."

He held out his hand.

"It's with the greatest reluctance that I leave you, Mr Blaise. I would much rather you would come with me at once."

"It's not the first time you have said it, Baynes."

"You and yours have been, and are to me, as my own folk. I believe you know that your happiness is mine—"

"The fates forefend! You know not what you are saying, man; it will be best for both of us if you never may. Now, Baynes, until the morning!"

"Until the morning, Mr Blaise, if you will have it so. Come to me as early as you can!"

"Yes, Baynes, I will come to you as early as I can."

As Mr Baynes left the room he paid no heed to a girl who stood at the entrance, clinging to the door-post as if she sought its support to help her to stand. Had he done so he could scarcely have helped being struck by the look of terror, of horror which was almost more than mortal, which was on her face.

The girl was Pollie Hills.

CHAPTER III

As soon as Mr Baynes had disappeared there was a chorus of voices. Men came crowding round the Gentleman, proffering congratulations. They were as excited as if the good tidings had been theirs. The deputy pushed through them.

"I'd have changed the five-pound note for you, Mr Polhurston; done it with pleasure, happy to do it. You should have took the fiver."

"What's a five-pound note to the Gentleman? He's got 'eaps more comin', p'r'aps 'undreds, p'r'aps thousands of 'em, for all that you can tell. I'm as glad, Gentleman, as if the luck was my own, straight!"

"I'm gladder! I 'ate to see a gentleman what's down on his luck, 'tain't natural. A gentleman always ought to have plenty of coin, so that 'e can be a gentleman; it's only right. As for me, I ain't a gentleman, and never sha'n't be."

"Here's luck to you, Gentleman! And may yer fortune bring you all the happiness you want, and something over, and may you never have another black hour—no, never, not as long as you live."

Through the little crowd pressed Pollie Hills, white-faced, trembling, eager. She caught him by the coat sleeve.

"Come, Gentleman! Now! At once! Quick!"

He perceived her with a smile. His face seemed to lighten at the sight of her.

"You, Pollie! Back! You've brought the sunshine! Bygones shall be bygones, eh? We've shared many a crustless hour; henceforward, together we'll quaff the flowing bowl."

She kept tugging at his sleeve, pulling towards the door.

"Come!—quick!—now!—now!—now!"

"Child, whence such hurry? Why in such haste to desert our friends?"

"Come! Don't I tell you to come? For Gawd's sake come! for Gawd's sake!"

She spoke as if she were half beside herself with fear and frenzy; as if her emotions were too insistent to permit of clear articulation. He made a little gesture to the others with his hand.

"One moment. We will return anon. The lady's business presses. Now, as ever, let it be our motto, ladies first!"

Together they went out, the girl urging him to a pace for which he seemed little disposed. The lad, Bob, watched their going with puzzled eyes.

Up the stairs, towards the street door, the girl went swiftly. At the street door she paused, holding him back with out-stretched palm.

"Wait! Let me see!" Thrusting out her head, she peered anxiously to the left and right. As if satisfied with the result of her inspection, again she caught him by the sleeve. "Come, now."

She drew him out into the street. In the badly-lighted alley it was nearly pitch-dark. The snow was falling; it lay two or three inches deep upon the pavement. It was bitterly cold. A keen north-easter blew through the lane. He shivered, refusing to advance, taking her by the wrist to prevent her pulling him.

"What's the matter with you? Why have you dragged me out of the warmth? Anything is better than out of doors on a night like this."

"Don't stand there talking. I'll tell you as we go. Come on."

She spoke with so strange a passion, and seemed, indeed, to be in such frenzied earnest that he yielded. He turned to go to the right; she wheeled him round as if he had been a child.

"This way, not that, for Gawd's sake!"

She hurried him through the lane, out into the open street. She herself seemed indifferent to the weather. Her shawl, unheeded, had slipped off one shoulder and was hanging down her back. He called her attention to the trailing garment.

"Gather your shawl about your shoulders, child. Are you in such a hurry, you must freeze?"

She did as he bade her without a word, tearing on as fast as she could walk. They passed a shop from which proceeded the smell of cooking. He stopped, bringing her to a standstill with a jerk.

"I don't know if you are aware that I am hungry, and that I've money. You must be hungry too. Is this business of yours of such pressing importance that you cannot spare a moment to permit me to buy us both some food?"

"Come on! Don't stop! For Gawd's sake come on!"

She glanced behind her affrightedly, as if she feared pursuit. Her mood affected him. He started off with her again at headlong speed, speaking, however, as he went.

"What is the meaning of this? Where are you taking me?"

"To Foster's Rents."

"To Foster's Rents! Why, in the name of goodness, are you taking me there? I know no one in Foster's Rents."

"But I do. I know a girl as has a room there. She's given me her key."

She showed that she was carrying a key in her hand. He pulled up, looking at her with amazement. He spoke with a touch of temper.

"And do you seriously desire me to understand that the meaning of all this to-do is simply that you wish to introduce me to the room of a young woman with whom I have no acquaintance, and desire none?"

"She's not there, she's out. She said I might have her room to-night."

"And pray, what difference do you suppose that makes? I have a shelter of my own. You have been rushing away from it as if death was at my heels."

"It is."

She spoke in a hoarse whisper. Her tone impressed him not agreeably.

"Death is at my heels! What on earth do you mean?"

"The traps are after you!"

She looked at him with a face which, in spite of the haste they had been making, was as white as the snow which fell about them. He echoed her words.

"The traps are after me?"

"Yes, the 'tecs! Come on, quick! I'll tell you all about it when we get to Sue's."

They renewed their break-neck speed. He perceived that the snow lay like a crown on her bare head.

"Cover your head with your shawl; you'll catch your death of cold."

She shook her head, as some wild creature might have done. Particles of snow flew in all directions. Then she hooded herself with her shawl. They continued in silence.

Presently she led the way through a narrow opening into another alley, glancing behind her as she turned. Low-built houses were on one side, a dead wall was on the other. She paused at a house which was about the centre of the row. The door was open. She entered the pitch-dark passage.

"Look out! here's the stairs; some of 'em's broken. You catch hold of my skirt, and then you'll be all right. Don't make no more noise than you can help." Stumbling up a flight of stairs, they reached the top. "There's six more steps here, then a doorway, and a sharp turn, so look out when you've counted six. I expect the door's open." The door was open, as she announced when she reached it. "Now turn to yer left; be careful how you turn; the passage is pretty narrow; it's only a step to Sue's."

He groped his way along a passage which was apparently well under three feet wide. Her voice came to him from the front.

"Here's Sue's." He heard a key being inserted in a key-hole. It was turned. "You go in first. I'll lock the door behind us as I come."

He went past her. He heard her withdraw the key, then, when she had also entered, insert it again, and lock the door inside. He stumbled against something; it fell on the bare floor with a clatter.

"The powers alone know where I am, or what I'm doing. The darkness is Egyptian."

"Have you got a match?"

"Not one. I gave the last I had to a gentleman in the street who possessed what I didn't—a pipe and a shred of tobacco. Under the circumstances it seemed superfluous."

"I daresay Sue's got some. I wonder where she keeps 'em. You'd better let me see if I can find 'em. I know this place better than you do. Perhaps they're on the mantelpiece."

She shuffled past him, brushing against him as she

went. She knocked something over, stifling an exclamation, then announced a discovery.

"I've found the mantelpiece." Something else fell to the floor—a small object. "I believe that's the matches; if it is, I've knocked 'em down." She went on to her knees to feel for them. Presently she struck a light. "I've got 'em."

Shading the light with her hand, she rose to her feet. A piece of candle was on the mantelshelf in a small white earthenware bedroom candlestick with a red rim. She lit it; the flame flickered for a second or two, then burned with some degree of brightness.

She turned to the Gentleman.

She was a handsome girl, with in her, it seemed, a strain of gipsy blood. The big eyes were Romany, the accentuated whites in such striking contrast to the velvety blacks. The olive cheeks were Romany, and the unburnished jet of the luxuriant hair. There was, too, in the supple poise of her body, in the lithe movements of her limbs, a reminiscence of the race which, for choice, had camped so long beneath the skies. The suggestion of the animal which was in the frightened attitude; of the wild creature of the woods, which is in sudden terror for its life, yet dare not fly; this also was Egyptian.

The man in front of her, with his glass, on its piece of string, stuck in his eye, and his fantastically shabby clothes gleaming with the snowflakes which he had not shaken off, and which, frozen hard, glistened like eerie gewgaws, seemed an odd companion for such as she. Some such thought, perhaps, occurred to him, for a smile wrinkled his lips, and a freakish twinkle was in his eyes.

They remained in silence for some seconds; she with her glance cast down, averted; he with his fixed on her, as if he sought to take her mental measure. Then he spoke.

" What did you mean by saying that the 'tecs were after me ? "

" They are."

" By the 'tecs do you mean the police ? "

" Course.'

" And, pray, what may these gentlemen happen to want with me ? "

" They know you done it."

" They know that I've done what ? "

" They know that you done for that bloke in the Embankment Chambers."

He was still. He did not move ; he did not let his eye-glass fall. Yet there was a quality in his silence which told he had been stricken, a fact of which she seemed to be aware, for a shudder went all over her. Then, with a leisurely uplifting of his eyebrow, he allowed his glass to drop ; and, taking an apology for a pocket-handkerchief from somewhere in the breast of his coat, he made a show of polishing its scratched surface.

" And, pray, what do you mean by saying that they know I done it ? You understand, I am asking what you mean ? "

" I told 'em."

This time he did move. He loosed his hold of his glass, and, with a half-step backward, clapped his hands to his side, as if suffering from a sudden twinge of pain. His voice was a trifle drier.

" You told them ? I thank you, child."

A sound came from her like the sob of a wounded animal. But she did not move ; nor did she change the direction of her down-cast glances.

" And may I inquire how you came in possession of the information which enabled you to assist these gentlemen ? I mean, how did you know I—done it ? "

" That there ring and bracelet you gave me was the bloke's. It says so in the bills."

"I see. And from that fact you drew your own conclusions. But you told me you had thrown the ring and bracelet both away."

"That was a lie."

"But I thought you never lied; that that virtue, at least, we had in common."

"That was a lie."

"I am sorry. It was a variation in a wrong direction. And where are these tell-tale trifles now? Hadn't you better hand them back to me?"

"The police have got them."

"The police?" His mouth shut tightly. It was a moment before he spoke again. "That is—that may be awkward. And why did you do these gentlemen this striking service?"

"Because I loved you."

"Because you loved me! The acuteness of a woman's logic!"

"And you used me bad."

"I used you bad? I shared with you all I had to share—all the good which, of late, the gods have given me."

"You turned me out into the streets."

"Not until I myself was turned; there, again, we shared."

"You said you never wished to look upon my face again."

"That was in a black hour. You will allow the moment was a trying one—I was so hungry, and so mortally ashamed. You, also, were a little trying."

"Then, to-night, I went to Benjamin's, and saw that bill, and about the murder, and the reward, and them things you gave me, and I knew you done it. Then I heard you outside a-saying—a-saying things, things that you meant that I should hear, and it made me clean wild, and—and I went off there and then and told 'em."

"I said those things because I was mad; this last day or two I've been very near to the place where madness lies. So it seems that, at the same time the devil entered into both of us. The coincidence is edifying. Do you know that, at this moment, I'm a man of fortune?"

"I heard the chap a-saying so."

"Yes, of fortune, positively, as it seems. And Baynes was good enough to hint that at last the night had passed and the sun of happiness had risen upon the horizon of the coming day. It doesn't promise to rise very high, upon my honour."

"Gawd help us both! You know I love you!"

For the first time her glances lifted. She threw her head back with graceful poise; her nostrils dilated; she looked at him with her splendid eyes, with, in them, the blaze of a great passion; her bosom rose and fell.

"I believe you do. It's very odd. Come here!"

He made a little gesture to her with his hand.

"I won't! I'm not fit that you should touch me! I'm only fit to drown!"

"That would be to make as poor a use of you as will possibly be made of me, and, after all, you are the finer animal. Well, we've had some pleasant times together, you and I—some rapturous hours, and even days—in which the world has been forgot, and it has been already heaven—in spite of certain disparities which might strike a casual observer as more capitally important than they really were. For experience has taught me that, where a man and woman are concerned, superficialities are not of much account; one returns to nature. I, too, as you have every reason to suppose, love you, though you've sold me for the pleasure of satisfying a moment's rage."

"Don't! Gentleman! Don't! Don't!"

C

Covering her face with her hands, her words came from her with shrieks of anguish; she shook as with ague. Replacing his eye-glass, he regarded her quizzically.

"Hush! Your strength is your weakness; apparently it bids fair to be mine too. Come, tell me exactly what it is you told these gentlemen."

She spoke from behind the shelter of her hands—breathlessly, without a pause; seeming, in her mind's eye, to be acting the part she had played all over again."

"I went to the station, and I saw a peeler, and I says, 'I want to see the inspector.' And they took me to a wicket, and there was the inspector looking through it out at me. 'I want to give imformation about the murder in Embankment Chambers.' 'What information do you want to give?' he says. 'I know the chap as done it.' 'What's your name?' he says. 'Pollie Hills.' 'How do you know he done it?' he says. ''Cause he gave me these, and I see from the bills as they're the things what was stolen from the bloke as was murdered.' Then I gave him the ring and the bracelet. When he caught a sight of 'em I saw him start. 'What's his name?' he says. 'Never you mind his name, I'll take you to him.' 'When?' he says. 'Now!' He puts on his cap, and goes to a door, and calls to someone. Then, all of a sudden, I saw what it was I'd done, and I gives a screech and I tore off as if the devi was after me. The peeler what I'd spoke to first, he tried to stop me. I give him a shove, and over he went, and I tore back to Benjamin's to give you the office."

"Do you think anyone saw which way you went?"

"The peeler went down pretty hard, and the inspector was taken by surprise, so before he could get to the door, I daresay I had turned down New Street, and was out of sight in Pleasant Place; I was moving."

. "So all they have to go upon is the ring and bracelet, and your name. You are sure you did not mention mine?"

"Certain. When they asked me, I said, 'Never you mind.' But as I turned into New Street, Aggy Mason was standing at the corner, close by the station door; she'd know me, and she knows that sometimes I put up at Benjamin's, and that's why I wanted to get you away."

"Do you think she'd tell them anything?"

"She might do; she's no particularf riend of mine, and never has been."

"Do you think that anyone who knew us saw us coming from Benjamin's here?"

"I couldn't say. There wasn't many about, was there? And those that were about I didn't stop to look at."

"How about the friend whose room this is?"

"She's all right. I met her early in the evening in the Cut. She says to me, 'Pollie, you're looking down on your luck.' I said, 'I am—I haven't even got the price of a doss,' which then I hadn't; another friend, what had a bit of luck, she lent it to me afterwards. Sue, she says, 'I'm going to spend the night with my sister over Greenwich way. You can have my room, if you like, and don't you hurry away in the morning, and then I'll come and have a talk with you, and I'll see if I can't put you on to something.' Then she gave me the key."

"You think she is the only person who has any reason to suppose that you are likely to be here?"

"So far as I know."

"And we shall be safe here for, at any rate, the next few hours?"

" I should think so."

" As safe, I suppose, as we shall be anywhere. Well, for present mercies let's be grateful. It's an odd sensation to feel that the police are hot upon your scent. Have you ever been in trouble, child ? "

" Never. Nor yet none of my people. We've all of us kept straight."

" Pray God you always may keep straight. I'm disposed to the opinion that this kind of trouble is the worst of all the kinds; and on the subject of trouble I claim to be a connoisseur. What's that ? "

The sound of voices from the street without was borne to them upon the wind.

" It's the people shouting."

" Yes; what is it they are shouting ? "

She went to the window, and opening it, leaned her head out. The voices immediately became more obvious.

" They are coming nearer."

" Fast. Can you hear what it is they're shouting ? "

" No."

" I can." He was standing at her side. Fitting his glass in his eye, he looked at her with his characteristically whimsical smile. " They're shouting, 'Hang him ! '"

CHAPTER IV

THE CHEST

SHE looked quickly at him, speaking as if an impediment was in her throat.

"They're not."

"I fancy you'll find that they are."

"How can you tell? You can't hear no words at all. They're just a-shouting."

"Listen."

They strained their ears. The hum of voices rose and fell. Each second they approached. But it was difficult to distinguish the purport of their cries.

"You can't hear nothing; they're just a-shouting."

"What's that?" At that instant a voice did rise above the rest, loud, strident, penetrative, vibrating, for a space, through the air. "Didn't that say, 'Hang him?'"

It was not easy to decide. Such might have been the interpretation of the yell. The suggestion having been made, one might tell oneself that it supplied the key. She withdrew a little from the window again all in a tremblement.

"What shall we do? What's happened?"

He gave his shoulders the familiar shrug.

"Who knows?"

"Do you think they're coming here?"

"Am I a seer?"

"Perhaps it's a fire."

37

"Perhaps."

She put her head again out of the window.   The sounds had increased in volume.

"They—they seem to be coming this way."

"They do seem to be coming this way."

"But they can't know I'm here, and anyhow they can't know you're with me."

"Stranger things have happened.   Give the police a clue, I should imagine them to be capable of following it up so as to arrive at a rapid and complete solution.   If they traced you to Benjamin's they would tell them there that we had gone away together, and all they would have to do would be to follow on our tracks, which I should suppose would not be difficult."

She turned to him with clenched fists and knitted brows.

"Look here, I'm all the evidence they've got.   If you was to kill me they'd have none.   You take and kill me!   I don't care how!   I'd love you to.   You kill me, because I love you!"

"How do you propose that I should kill you?"

"Put your fingers round my throat and choke the life right out of me; it's easy done.   I'll keep quite still.   I'd love to die with the feel of your fingers round my throat.   Gentleman, do!"

"I'm of opinion that the remedy would be worse than the disease."

"Then I'll kill myself; I'll do it quick.   I'll jump over into the street.   I should think that'll do it if I fall on my head on the stones."

She began to scramble through the window.   He caught her by the waist.

"Hasten slowly.   I would rather, if you have no objection, that you should not kill yourself, and that I should not kill you.   I would prefer that you should continue to live, even though you love me.

Your death, in any case, would hardly work for my salvation. You've given them the clue; with or without you they'll be able to follow it home."

"Do you mean I've hung you, anyhow?"

He smiled, that smile of his which suggested a comic appreciation of the situation rather than the enjoyment of a particular jest.

"I don't think I'm hung as yet."

She made a movement towards him, which was marked by that impetuosity which seemed to be a characteristic of all she said and did.

"Kiss me!"

She raised her lips to him. He touched them with his own. Throwing her arms about him, she strained him to her, kissing him as though she sought to draw all the virtue that was in him through his lips.

"I love you!"

She released his lips to breathe the words in a sort of ecstasy. With one hand he smoothed her tangled hair, smiling at her all the time.

"I believe you do. I am blessed above my fellows. Love and fortune, the two things most ardently to be desired. What could a man want more, with those voices in the street?"

She drew herself away from him.

"Perhaps—perhaps they're not coming here."

"Perhaps not."

As if to mock them, the voices rose in sudden clamour, louder than anything which had gone before, and nearer. She made for the window.

"I can't see anything. I don't know what's the good of a window when you can't see any more out of it than you can out of this. I'll get outside and peep over the edge.

The room in which they were was an attic. The window, which was a small one, was some two or three feet from the front of the house. The roof sloped down to a parapet, which could be dimly seen through the falling snow. It was down to the parapet she proposed to climb, so that, by peering over the edge, she might obtain a glimpse of what was taking place in the street.

He stopped her as, with a spring, she raised herself on to a wooden window sill.

"Promise that you won't throw yourself over."

"I'll take my davy that I won't, not if you don't want me to."

"I do not want you to. Also be careful that you don't slip."

"I sha'n't slip, I'm as surefooted as a cat, I am. When I like I can stick to the side of a wall like a fly."

He let her go. She allowed herself to slide down the snow-covered roof till her feet stayed against the parapet. Then cautiously she bent herself forward till the roadway beneath was within her line of vision. The din had become surprisingly great. Men, women, children, were shouting against each other, not apparently in anger, but in the enjoyment of some huge, common joke. Even yet it was not easy to distinguish what it was they said, the voices were so hoarse, the discords so confusing.

The girl, twisting round, extending her arm, catching at the sill with her hand, drew herself up on her stomach. The Gentleman was peeping out, so that their faces met.

"What has appealed to the public's sense of humour?"

"It's the peelers."

"The peelers?"

"Yes, there's a lot of them, I can hear their feet a-keeping step." She returned to her post of vantage, looking round now and then, to shoot at him short sentences which were pregnant with meaning.

"They're at the end of the street. The crowd's a-guying 'em. There's a row; the peelers are trying to keep 'em out of the street, and they're trying to get in."

That there was what she termed a "row" was obvious; if one had ears one scarcely needed eyes. That one set of persons was trying to do something which another set sought to prevent their doing was palpable to the partially deaf; also, that the first set was a noun of multitude.

Again the girl drew herself up to the window.

"It's them! They're on to us! What shall we do?"

"Wait here and allow myself to be taken like a rat in a trap, or shall I go down to them into the street and say, 'Behold the man?'"

"That you sha'n't."

She spoke between her teeth. Her eyes were blazing. An authoritative voice was heard in the street below, issuing orders. There was the measured tread of disciplined feet, which halted. Suddenly a woman's voice rose up, apparently from immediately in front of the house. She was bawling,—

"I'll show her! Taking—murderers home to my —place!—her!"

She concluded with a volley of oaths. The girl, still lying face foremost on the roof, started. She caught her breath.

"That's Sue! She's been drinking."

"There is a suggestion in the lady's voice of a recent conscientious endeavour to assuage a copious thirst."

The woman in the street resumed her observations.

"That's my house, Mr Copper! That's my place! That's my room, up on top there, to which I give her the key. I said to her, she being a friend of mine, 'Pollie, you take the key, and you go to sleep

at my place.'   But if I thought she was going to take
a somethinged murderer home with her, I'd have seen
her somethinged first."

The girl on the roof gasped.

"Sue's given us away, that's what it is.   That's
the worst of her, she drinks, and then she's not the
same.   She wouldn't have done it if she hadn't been
drinking; she'll be sorry in the morning."

"It seems to be a feminine attribute to be sorry
afterwards.   So I presume that nothing remains for
me but to surrender with as great a show of dignity
as is within the range of my capacity."

"You won't do nothing of the kind.   I'll be even
with Sue, now I know it's her.   I'm not going to be
beat by Sue.   You go and take the key out of the
door, and blow the candle out, then come outside to
me.   It's not all over yet; I got you into the hole,
and I'm going to get you out of it."

He made a little grimace, then did as she suggested, as quickly and quietly as he could.   He gave
her the key, then clambered out to her on the roof,
she whispering instructions as he came.

"Take care how you show yourself; don't get up
too much, or they'll see you down below."   She
closed the window after him.   "Now, who's to know
we've been in the room?   There's no key in the door,
no candle lit, the window shut, what's to show we've
been there?   The peelers 'll think Sue brought 'em on
a wild-goose chase, and they'll give her a rare old
wigging.   Serve her right if they do!"

"And is it part of your plan, then, that we shall
spend the night out here, on the principle that it's
better to be frozen than hung? for frozen we shall be
if we stay—stiff."

"It is coldish."

Her teeth chattered as she spoke.   They felt the
wind more up there; it blew the snow into their

faces in blinding whirls. The temperature seemed to be falling as the night went on. Her companion was seized with a paroxysm of shivering.

"It's coldish, as you say. It's cold, not fear, which makes me shake; it's getting into the very marrow of my bones. I would remind you that hunger is not the best preservative against a chill."

She was listening at the window.

"I believe they're knocking at the door."

"Then we'd best be moving. They won't be long before they're through it, and when they find that we're not there they'll look out of the window, and then we shall gain nothing by the chill we've caught."

"Do you think you could walk along the roof? Down by the bit of wall it isn't difficult, only you must stoop low down so as they sha'n't see you from the street.',

"I'll try. I thought I'd gone through the whole gamut of the adventurous, but I've come on a new chord to-night."

"Come on, then. Keep close behind me, and stoop low down, only, for Gawd's sake, don't slip."

She led the way. The parapet took the form of a low wall, perhaps two feet high, so that by keeping close to it, and bending nearly double, aided by the darkness and the snow, their movements were hardly likely to be seen from the street. They had gone some little distance when he pulled at her skirt. She stopped.

"What's up?"

"We're leaving our tracks. They've only got to look out to see that we've been here."

"That's what I was thinking. We'd better get into one of these other houses, and try our luck that way. It's about our only chance. We'll try this one. There don't seem to be no light in the room.

Perhaps it's empty, Lord send it!" They had passed by a window through which there shone no glimmer of a light. "It's no use our trying to drop off, the peelers is sure to be all round the building. If we didn't break our necks we should fall into their arms. I'll go right on to the end, so they may think we've tried it on, and then I'll come back to you. You wait here, I'll be as quick as I can.

She was quick as she could be, scurrying like a cat, under cover of the parapet, along the whole block of buildings, then swiftly back again. She found him listening at the pane.

"Do you think there's anyone inside?"

"I haven't heard a sound."

"Have you tried the window?"

"It's one of those which swing on hinges. It's hasped inside. We shall have to push it open."

"Go on, push it, then."

"I can't. I seem paralysed with cold; my strength's all gone."

"Gentleman!" It was an interjection, uttered with a catch in her throat. "Let me come!"

Putting her shoulder against the casement, she gave it a vigorous shove. The rotten fastenings yielded; the window was open. Not a sound proceeded from within. She commented on the silence.

"There don't seem anyone inside to ask us who we are and what we want, and it's no time to be asking leave. I'll go in first."

When it came to his turn she had to render him assistance to enable him to enter. As he said, he seemed to be paralysed with cold. When they were both in, she closed the window, holding it to with her hand. All continued still. They could hear each other breathing.

"There's no one in here but us, so that's all right." Yet she spoke in a whisper, as if fearful that there might be someone close at hand to overhear. "I wish we'd brought a match from Sue's. This darkness isn't just the thing when you don't know where you are."

"I have. I took the box when I blew out the candle. I've been carying it in my hand. I'm afraid it's wet."

"Never mind, I'll manage. Thank goodness that you brought it."

· The first she tried refused to ignite; the second burst into a flame. With its flickering aid they began to examine their surroundings.

"There's no one to be seen, at any rate. It seems to be a queer sort of crib." It did; more like a cabin on board a ship than a room in any dwelling-house. A hammock was slung in a corner. The furniture was not only heterogeneous, it was unusual. On all sides were recollections of the sea. "What's that horrid-looking thing up there?"

"I fancy that's a stuffed alligator. The rightful occupant of the room seems to have a taste for zoology. That's a monkey over the door, and a baboon over the hammock; there's birds and beasts of all sorts and kinds."

The match went out. She lit a third. Her eyes went wandering round the room. All at once she gave an exclamation.

"My crikey! I believe I've got it!"

"Got what?"

In her excitement she jerked the match out. She lit a fourth.

"What's the betting that I've not got it? If those peelers 'll only give us time we'll diddle 'em yet, and we sha'n't want much time either."

"A little more lucidity on your part might assist my torpid wits."

"You see this?" She was standing before a

wooden chest of unusual dimensions. It occupied nearly the whole of one side of the room. It was probably about seven feet in length, three in width, and two in height. How it had ever been brought into the room was a mystery. It was painted green. On the top were books and papers, odds and ends. "Here, catch hold of the matches. Light one at mine. We'll have a look at what's inside ; Lord send it's empty. Drat that window, it's blown open! I'll have a try at wedging it."

Taking a paper off the lid of the box she succeeded in inducing it to act as a wedge. The window remained closed. She returned to the chest, at which the Gentleman was peering by the light of the match he held. With much celerity she removed the books from off the lid.

"Somebody's left the key in the lock, all nice and handy." Turning the key, she raised the lid. "Here's a stroke of luck—it's empty. Now you listen to me. You see these holes?" She referred to three or four rows of holes which were bored in front of the chest, from end to end. "Air's bound to come through these—plenty of it. I'll lock it up and hide the key. There's some tools in the corner; I'll give you a hammer, a chisel and saw, then you'll be able to make your own way out when the coast's clear."

"Will you be so good as to explain what it is you mean in somewhat greater detail? You seem to think I ought to understand you, but I don't."

His tone was dry.

"It's plain enough. You get into the chest, I lock it, I give you the tools, as I say, and then, when the coast's clear, you use 'em to get out. With them to help you you'll be able to get out as easy as winking."

"When the coast is clear! Do you suppose that this chest won't be the first object to catch a policeman's eye, and that he won't insist on seeing what is in it? Your hiding the key will only serve to whet his curiosity."

"That's only half the game, your getting in here. I'm going to be you!"

Her tone was positively buoyant; the immediate prospect of doing something desperate seemed to have raised her spirits.

"You are going to be me? I wish I could be you!"

By the radiance of another match she pointed to some clothes which hung on pegs in a corner—masculine habiliments.

"You see those? I'm going to get into those togs, and lay out that I'm a man, that I'm you. When those peelers do come, I'll give 'em such a handling they'll make sure I'm you. I'll treat them to such a scrapping match they'll be glad enough to make a capture, and won't stop to ask no questions. They'll haul me off, and themselves along with me; then'll be the time for you to do a bolt."

"Child, do you suppose—?"

"I don't suppose nothing. You're going to get in, and that's all about it. Here's the tools, there's a chisel, hammer and saw, and a mallet, in case you want it. Now in you go. I got you into this hole, and I'm going to get you out of it, only every moment is precious."

"Pollie!"

"Gentleman!"

He seemed to be overcome by the stress of his emotions—by hunger, weakness, cold; he was trembling like a child. She threw her arms about his neck, and kissed him—once, twice, thrice. Then, though he was taller than herself, she lifted him off his feet, and actually laid him in the box, he appearing incapable of offering even the least resistance. She fell on her knees beside him, and, stooping over, kissed him once again.

"Gentleman," she said, "Gawd help us both!"

With that she shut down the lid of the chest.

CHAPTER V

THE CHASE

DOUBTS came to her so soon as the box was shut. She hesitated, with her fingers on the key. An idea occurred to her. The match she held was burning nearly to the end. She passed the flaming fragment to and fro before the holes which were in the side of the box.

"Gentleman," she said, speaking with her lips close to the holes; "Gentleman, do you see the light?" No reply. She called again. "Gentleman!" Still not a sound from within. "Gentleman! don't you hear? Speak! tell me, don't you see the light?"

She rapped with her knuckles against the wood. Still there was no answer. The flame reached her finger tips. With an exclamation, she dropped the match. It went out. She raised her finger to her lips.

"By gum! that give me beans!" Raising the lid of the box, she spoke once more to the man within. "Gentleman!"

No response. Her ear, keenly alert, caught a sound from without.

"What's that? On the roof! They're coming!"

Shutting the lid with a little bang, she locked it and withdrew the key.

"What shall I do with it?"

D

She rose quickly to her feet. Lifting one end of the box, she slipped the key beneath it. In the dark she stood listening.

"They're coming! They've found our tracks, and they've got on to the roof to see where they lead to. I wonder if I shall have time."

She began to tear, rather than to take off her clothes, dragging them from their places with eager, flying fingers, with complete disregard to any injury they might receive.

"I wonder what's come to the Gentleman? Why he don't speak to me? if I done wrong in putting him in the box? Gawd help us both, there ain't no time to find out! I'm in luck if I can get into them other things before they're here! What's that?" She stopped to listen. "They're passing the window —that's one; that's two; that's three; and that's all." She held her breath before she made sure. "There's three of them! They're following my tracks to see where they lead. Good luck go with 'em! I wish they'd throw themselves clean over! Now for them other togs. What shall I do with these?"

She had stripped herself nearly to the skin. As best she could, for the darkness, she was gathering up her discarded clothing from off the floor.

"I wish I dared to strike a light! How am I to see where to put these things? If they catch sight of them the game's all up. I shall have to chance it, and chuck 'em behind the box."

She did as she said; with difficulty, in her haste, finding room for them between the chest and the wall. As she crammed them anyhow away, she struck the lid smartly.

"Gentleman! Gentleman! Speak to me." Her appeal went unnoticed. There was a catching in her breath. "I wish I knew what's gone wrong with him, I wish to goodness that I did! I wonder if it would be safe?"

She considered, shivering, in her single garment, with the cold.

"Better not, I might only be giving him away."

She felt for the masculine habiliments which she had seen hanging on the pegs. Pulling one of them down, she endeavoured, by the sense of touch, to discover what it was.

"What's this? It's trousers, that's what it is. Which is the way in? Seems as if this was; anyhow, here goes. What's this dangling behind? Braces! That's a bit of all right; I don't know how I should have kept them up without 'em. Now for the next thing, please."

The next thing proved to be a waistcoat, and that was followed by a coat; both were sufficiently capacious.

"I can't say much for the fit; there's room for another one along o' me inside of this, and perhaps a trifle over; but that don't make no odds, it isn't as though I was going to have my photograph took; if I was, I would come out a treat. I'd like to take a squint to see what I look like before these nice young gentlemen came. Where are they? Not fallen over into the street? No such luck! A copper never hurts hisself; he couldn't do it, not even if he was to try. I wonder if I'd have time?" Again she rapped on the lid of the chest. "Gentleman!" Again no answer. "I wonder what has gone wrong with him? My Gawd! if I only knew I'd have a couple of minutes—!"

Her ear caught a sound.

"What's that? They're coming back! It's no good, I'd better take no chances."

She stooped down over the chest, speaking as loudly as she dared, so as to make sure of her being heard by the man who was inside.

"Gentleman! bear up! If you're feeling a bit queer, never you mind, you'll be better soon! Don't you worry, I'll see you through. Then, when you've the chance, you get clean away. Don't bother about me, I'll be all right. Good-bye!" A pause. "Gentleman! Good-bye!" Another pause. "Say good-bye to me!" Still no answer; again that catching in the throat. "I'd wish he'd say good-bye to me; but if he won't I suppose he won't. Now for them nice young coppers."

She listened, moving towards the window.

"They're taking their time in coming back. I wonder what their little caper is? If I had gammoned 'em, and made 'em think that we had dropped over the wall at the end, what a stroke of luck that'd be. Too good to be true, a lot too good." There was a noise outside. "Hollo! That doesn't sound as if much luck was going to come my way."

She went close up to the window, straining every nerve to listen. Voices were heard without.

"Seems as if the snow had been trampled down pretty well just here, as if someone had been sitting down in front of this window, or kneeling down."

"It does seem like it. We'll have a look."

On a sudden a blaze of light came through the casement, throwing the fairy-like tracings made by snow and ice upon the glass into mysterious radiance. It gleamed on to the girl's face, and into her eyes. In an instant she had dropped on to her knees, but not quickly enough to be able to be quite sure that she had escaped observation.

"The bull's-eye! Darn 'em! I'd forgot it. How it does shine. I wonder if they caught a sight of me?"

Again the voices were audible without; she holding her breath, so that nothing might escape her.

Apparently someone was subjecting the window to close examination, with suspicions already aroused.

"It strikes me that this window's been opened, and that not so very long ago; here's the snow fallen away all round it."

"See if there's anyone inside."

Knuckles rapped against the glass; there came a sharp official inquiry.

" Who's inside there? "

The girl whispered to herself an answer.

"Yes, who is inside? If you find out, perhaps you'll know."

The bull's-eye lantern was moved hither and thither, lighting up rapidly, one after the other, every corner of the room. The singularity of some of the objects on which it gleamed apparently struck the searcher.

"Queer shop in there; seems a sort of museum; there's a lot of stuffed animals."

"Yes, and there's one that isn't stuffed waiting for you."

The shafts of light, travelling here and there like will-o'-the-wisps, passed over the girl, who was crouched close to the wall immediately under the window. Another voice was heard—that of an official superior.

" Who lives in here? "

Apparently the proceedings had attracted the attention of the occupants of other attics down the row. It seemed tolerably obvious that, with heads protruded from their windows, they were watching what was taking place with interest and satisfaction. A reply came to the inquiry—as it seemed to the listening girl, from a pair of feminine lips belonging to a head which, more than probably, was thrust out of the very next window, the one on her left.

"Bill the Sailor, he lives in there; he's mostly

down at the wharves, and he's hardly ever home
at nights. I don't suppose there's no one in there,
not now there isn't, leastways no one as ought to
be there."

"Have you heard anyone moving about? I suppose you can hear from your place?"

"Oh, yes, I can hear right enough, especially if
there's anyone what's at all noisy; the walls ain't
so thick as all that comes to; but I ain't been in
long myself, and I ain't heard nothing since I have
been in."

"See if the window's shut."

A hand was pressed against it, shaking the frame.

"I don't believe it's hasped; it seems to be stuck
at the bottom."

"Drive it open."

The girl within drew a long breath.

"Now's the time to begin the game."

She stood up straight, directly in front of the
window. For an appreciable space of time she was
still, possibly to give herself an opportunity to get
the better of the quivering something which went
right through her, from the tops of her fingers to the
tips of her toes. Then she said, in a voice which
was so like a man's that nobody, who was not in
the secret, could have helped being deceived by it,—

"I'm in here—that's who's in here; and now you've
got it. And the first copper who puts his head inside
that window 'll get it broken; that's another little
bit of news I'll give you cheap."

The sudden up-speaking of the voice, whose tones
were not so much loud as significant, produced an
evident effect on those without. The attack on the
window ceased. There were hurried whisperings.
Then a stern inquiry.

"Who are you?"

"Never mind who I am; though, if it'll do you

good to know, I don't mind telling you that I'm a gentleman by birth and education, what's ready to smash the jaws of any dozen dirty coppers as ever mistook themselves for men. Break it gently to your mothers."

"Open the window!"

"Not me! I'm not going to open any window, not hardly, a night like this. And don't you open any window either, or it'll be the last one you will open, just for a little while."

There was the movement of a lamp. Like lightning the girl, ducking, sprang against the wall, hugging it as closely as possible, keeping well within reach of the window. The blaze of a lantern came through the pane, traversing the room in all directions. The first lantern was joined by another, the two shafts of light crossing and re-crossing each other so rapidly as to render it difficult for any movement within to remain concealed. Pollie thanked her stars that the Gentleman was out of sight. She watched the splashes of golden colour flashing up and down, from side to side, with an increasing sense of satisfaction. The blood in her veins was growing warmer.

"Can you see him?"

"No. He must be against the wall."

"Slant your lanterns."

The girl dropped on to her knees. She bowed her head. The light passed over her, revealing nothing of her whereabouts, though she seemed to feel it touch her as it went. It made her pulses tingle, her eyes gleam.

A voice entered with the flashing gleams, as a sort of accompaniment.

"We know who you are, my man, and all about you; so, if you take my advice, you won't give us any more trouble. If you'll open the window and surrender quietly we'll be gentle with you. If you won't you must expect what you'll get. You've given us enough trouble already. Do you hear?"

The girl did hear. But she did not say so. Instead she crept a little closer to the window. With both her hands she picked up something which was lying on the floor, something which seemed to her to be a board, of a useful size and weight. Her instinct was not at fault. Hardly had she possessed herself of a possible weapon than the casement was flung open. There was an instant's pause, as if those without were awaiting developments from within. Then a head was intruded, one which was crowned by a helmet.

"Now then!" said a voice, in a tone which was perhaps meant to convey, at one and the same time, both a warning and advice.

But that was all it did say. When it had said so much, the girl, rising to her feet, brought her weapon down with a crash upon what she could see of the intrusive object. The blow knocked off the helmet, which struck her feet as it fell to the floor; the head was withdrawn more rapidly than it had entered. Quick as thought she pushed the window to, re-wedging it. A colloquy took place without, which was audible within.

"Has he hurt you?"

"Stunned me pretty well."

"What did he hit you with?"

"It seemed to me like a bar of iron. I know it cut clean through my helmet."

"Did you see him?"

"Saw nothing. He downed me before I had a chance."

There was a note of grievance in the speaker's tone which struck Pollie as diverting. She laughed right out. The sound, penetrating the pane, seemed to strike those without as a little ill-timed; at any

rate, judging from the tone of the voice which commented on the burst of merriment, it did not tend to
increase their gaiety.

"All right, my lad, laugh away! We'll give you
something to laugh at before we've done with
you."

There was silence; or, if anything was being said,
it was too low to reach the attentive ears inside.
Possibly they were examining the wounds of the
injured man, or discussing what it would be best to
do next. All at once a voice cleft the night-black
air, proceeding, apparently, from an attic window
some three or four doors off—a stentorian voice, which
rose above the screaming of the wind, a voice in
which there was the delight of battle, a strain of
jubilation.

"They've found 'im! 'E's broke a blimey copper's
'ead, and give 'im what for! Cheer, boys, cheer!
'E's a game un! There's goin' to be skittles, I give
you my word!"

The solo was followed by a chorus. Shouts arose
from all sides. One's ears were saluted by remarks
which, taken in the mass, could scarcely be regarded
as conveying compliments to the officers of the law.
Of all the things which were wished them, not one
was in any way to be desired. Apparently some of
the choristers were meditating a passage from words
to deeds, for presently an official voice rang out.

"Now, don't any of you people get outside them
windows, there'll be trouble if you do. You get inside there! Do you hear? Keep inside your own
rooms."

There were jeers in response, the stentorian voice
roaring above the rest.

"Who's getting outside? And what's it got to do
with you if I am? This is my own window, I suppose, and what's it got to do with you if I choose to

get outside of it to take the air? Who pays the rent —is it you or is it me?"

The official voice replied, with that mixture of authority and good-nature which, even under the most trying circumstances, is not the least striking characteristic of the London constable,—

"You shut your head, my man, and take care you don't catch a bad cold. You know what you've got to do; take my advice and do it."

The stentorian voice continued the discussion, but the officer paid no heed. He called over the parapet to someone down in the street below.

"Our man's in here, in the front room at the top of the house. The window's not easy to enter with him inside. Send in two or three men, and tell them to break down the door, if necessary. We'll stop up here."

The girl inside could hear that the order was being obeyed. Footsteps were distinctly audible entering the house, ascending, with heavy, measured tread, the stairs. Higher they came, and higher; nearer and nearer. She grasped more tightly her weapon of offence. Someone spoke, someone whose voice she had not heard before.

"Look out there; careful how you go; these stairs aren't easy. Be ready for a rush."

"A rush." The girl smiled, as she murmured to herself, "I'll give you rush enough in half a jiff, my beauties!"

The handle was tried—once, twice, and yet again. Then there was a rapping at the panels.

"Inside there! Open the door!"

Again the girl feigned to speak as a man, and again not loudly, but rather as if what was said was meant—and even more.

"I'll see you the other side of Jordan first. I'm not going to open the door, and I'll kill the man who does."

Even as she spoke she was all in a palpitation—at that last moment she had remembered her hair. If anything like a scrimmage took place it would inevitably come down; her sex would be revealed; her attempt at deception at an end. The simplest policeman could scarcely be expected to believe in a man whose hair hung down below his waist.

"To think I should have forgotten! What a fool I am! That's just like me! If the game should be spoilt because of my hair!"

She clutched at her locks with angry fingers. But, despite her agitation she kept her head. The light, still flashing through the pane, rested for a moment on the mantelpiece. It shone on what looked, in the momentary glimpse she caught of it, like a knife cased in a leather sheath, hanging from a nail just above the shelf. The light passed. Scurrying towards the fireplace, under cover of the wall, she snatched at the object whose presence the unfriendly lantern had revealed. It was a knife, and a sharp one; it cut her to the bone when she pressed her finger to the edge, after she had drawn it from the sheath. Grasping "the glory of a woman" with her left hand, holding the knife in her right, she severed the entire mass as close to her scalp as she could. Kneeling, she thrust the ravished locks into the fireplace, and up the chimney.

All the time the voice was addressing her from without words of counsel. The speaker was alluding to her threat to kill the man who opened the door.

"Don't you try to do anything like that, my lad. Be reasonable, don't you be so silly. There are too many of us, and you'll only make it worse for yourself."

With such rapidity had she performed the process of shaving and shearing that already by the time the voice was still, the deed was done. She breathed defiance in reply, while she thrust the hair up the chimney.

"I'll make it as bad as I can for you, you can take my word for that, Mr Copper. Don't you trouble your head about mine, you look after your own; if I can bash out your brains I'll be content, my bloke."

By the time she had made an end of speaking she had deposited, to the best of her ability, her hair in its hiding-place. She threw herself flat on the floor, to dodge the lantern which still inquired for her whereabouts.

The voice without continued.

"If that's your style, my lad, it'll have to be. It's as you please, only, don't say when it's too late I didn't give you fair and proper warning. Now, men, close up. Force the door; then rush in all together."

At the first assault the door yielded, showing the strength of its fastenings. It flew open as if it had been secured by nothing stronger than cotton threads. The speaker without proved himself to be a better tactician than his closing words suggested. He did not urge his men to make a blind rush forward as soon as the barrier was no longer in existence. On the contrary, he held them back, as if he preferred that the rush should come from within. There was a pause, as if each side was waiting for the other to make the opening move; a pause which was broken by the man at the door.

"Foster!" he cried.

A voice came from the window.

"I'm here. Is that you, Wilson?"

"Yes, I've got the door open. When I give the word, you come in through the window and we'll come in through the door."

"Right!"

"Now!"

CHAPTER VI

THE CAPTURE

THE casement was driven back with a crash against the wall; if it was not torn from its hinges it was because they were very much stronger than had been the fastenings of the door. While the girl still hesitated, a dark figure, clambering through the narrow window, sprang on to the uncarpeted boards. It was followed by another.

"Show a light out there!" exclaimed a voice.

In response, lanterns flashed through the window and from the open doorway. Their appearance gave her an idea. She still grasped the knife which she had applied with such effect to her whilom luxuriant locks. For one moment she contemplated using it against her assailants; but even that brief space for reflection was enough to show her that to do so would be to make the desperate game which she was playing too expensive both for herself and them. She was crouching in the shadow of a curiously-fashioned wooden chair. Laying the knife down on the floor beside her, so that she might be safe from the temptation of using it in some moment of frenzy, her fingers brushed against something on their passage to the ground. It was the policeman's helmet. Snatching it up, thrusting it on her head, she passed the strap beneath her chin. Gripping the

weapon of offence, whose capacities she had already
tested on the intruding head, flitting noiselessly, as
nearly as possible on all fours, she reached the door.
Lifting her weapon, before they realised that she was
upon them she brought it down, crash! crash! upon
the two lanterns which their owners were using as
search-lights. So unlooked-for was the assault, and
so well-directed were the blows, that before the officers
knew what had actually taken place, the lanterns
were struck clean out of their hands, and fell
clattering to the floor.

Chaos followed ; even policemen sometimes lose
their heads.

"There he is!"

"Look out!"

"On to him!"

"Down him!"

Each officer proffered a suggestion of his own.

In the confusion occasioned by the sudden dis-
appearence of the lights, someone did strike at her
a blow. Its bourne was the already sufficiently ill-
used helmet, which, in spite of the detaining strap,
it dislodged from her head. What she had foreseen
might happen, did happen. The outraged constables,
demoralised by the disappearance of their lanterns,
made a hasty dash into the room, with the intent to
close, on any terms, with their unseen foe. The door
was left unguarded. Whereupon the girl, crouching
on the floor, slipped past their legs, and like a flash
she was out of the room.

Confusion became confounded. Those in front
scarcely knew what had happened. It was the man
outside the window, with the lantern, who detected
the flight.

"He's gone out of the room! After him! Why
the devil did you let him go?"

Down the stairs sped Pollie. At the foot of the

first flight stood a woman, who, with arms akimbo, was evidently appreciating the proceedings to the full. She was apparently the occupant of an apartment, whose door stood invitingly open. This quick-witted lady grasped the situation in a second. With a movement of her head she directed the fugitive towards the open door.

"In with you! I'll keep 'em out as long as I can. The window opens on to the mews; it's nothing of a drop."

Through the door dashed Polly. The woman closed it after her, remaining on the threshold, with her back against it. To her came Policeman No. 1. He did not move so swiftly as the quarry; he was portly, and no longer a lad.

"Which way did he go?"

"Down the stairs," replied the woman, with the most complete *insouciance*. "He's out into the street by now."

Policeman No. 2 interposed with a point-blank contradiction.

"That's a lie! I heard you speak to him, and I heard you close your door. He's in your room."

The lady proved herself a virago.

"In my room!" she screamed. "What do you mean by that, you dirty-faced monkey! And me a respectable married woman what's buried two husbands and seven children!"

Policeman No. 1 treated her with scant ceremony.

"Get out of the way."

He pushed her aside, whereat she attacked him tooth and nail, giving him as much as he could do to preserve himself from being torn to shreds. His colleagues, instead of coming to his assistance, took advantage of his preoccupation to enter the lady's premises. There the open window told its own tale, which received corroboration from the cries without.

Pollie, acting on the woman's hint, rushed to the window, threw it open, and, without pausing for consideration, lowered herself till she hung by her hands to the sill, and then, trusting to fortune, dropped into space. Space, however, in this case was but a figure of speech; her feet touched ground almost as soon as she let go. Gathering herself together to start afresh, she found herself gripped tightly from behind. She had dropped, to all intents and purposes, right into a policeman's arms.

A head was protruded from the window above.

"Got him?"

"Yes, I've got him."

But, in so asserting, the speaker was slightly premature. His afterthought was better advised.

"Some of you chaps had better come down and lend me a helping hand."

The words came from him in gasps. The efforts he was compelled to put forth were such that it was strange that he could speak at all. His captive seemed to be possessed of the resources and the agility of a wild cat. In no way could he get a firm grip. Just as he thought he was succeeding, he lost his footing on the slippery snow, and in another moment he was seeing stars upon his back, with the fugitive once more in headlong flight. A row of black-coated figures, dropping, one after the other, from the window, alighted at the chagrined constable's side; they were his colleagues—arrived a little late.

Over the snow tore Pollie, the officers sprinting against each other in the rear. A figure was advancing from the front—another constable. Gritting her teeth, the girl rushed furiously on. In a few more seconds there would have been a collision, with dire consequences, no doubt, to Pollie, when something induced her to swerve from her course. A man was holding open a stable door in a fashion which was, to say the least, suggestive. She took it as an invitation on which it might be as well to act. She dashed past it, the man shut it after her, locked it, and pocketed the key.

Up came the excited constable.

"You open that door?"

"What door?"

"You know very well what door I mean! Do you think I didn't see you let him in?"

"Let who in?"

The man's innocence was sublime. Other constables had arrived upon the scene. They all made him the object of their attentions.

"None of your nonsense! Who do you think you're playing with? You hand over that key."

They all evinced a disposition to treat him with but slight consideration.

"What key?"

He took something from his pocket, flinging it right over the roofs, high into the air. It was the key of the stable door. The action seemed to increase the policemen's agitation.

"All right, my man! This is the worst night's job you've ever done in all you're life. You shall pay for this."

"Pay for what? Can't a man do as he likes with his own door key? Things is coming to a pretty pass if he can't chuck it away if he's got a fancy to. What next? I never see nothing like you coppers."

His observations were suffered, for the present, to go unheeded. Instructions were rapidly issued.

"One of you men—you, Wilson—go and tell them to keep a look-out in Duncan Street; these premises look out on their back yards. As fast as you can!'

A policeman went racing over the snow-covered cobble-stones. Others had been examining the construction of the stable door.

E

"This is a solidly-built door, this is. It will take some time to force it—unless we can pick the lock.'

"You won't pick that lock in a hurry, even if we'd got the tools, which we haven't. Better try the private door."

They tried the handle of the private door, which refused to yield. They turned to the tenant of the place,—

"Where's the key?"

"Where's what key? The key into my 'ouse? Haven't got one—never have had one—don't use one. When I go in, I go in through the stable, and that's the way you'll have to go in if you must make a burglarious entrance into a gentleman's premises."

"Any of you know anything about this man?"

"I do, sir. His name's James — Bandy-legged James, they call him. Drives a four-wheeled cab. He's a bachelor; lives here by himself; there's no one else upon the premises."

"Then, in that case, it will be no good ringing. Force open the door!"

A stalwart constable dashed his shoulder hard against it.

"I can do it," he announced. "It's only on a single bolt."

While the assault was being made, the tenant commented on the proceedings in a fashion of his own.

"That's right! Smash the door down! Never mind what I think! Break my 'ome up for me! You're a pretty set, you coppers! Talk about you being our protectors—as I see it in the paper! You look as if you was my protectors, don't yer? By Gawd, I'd like to break into your 'omes, like you're breaking into mine!"

By this time the door had yielded. The chief officer examined the lay of the land; a flight of stairs was in front, a door on the left.

"This door leads into the stable; it's locked. If it's been locked all the time, then perhaps he's still inside. Force it open!"

This time the operation was the work of a moment; the door might have been made of matchboard, so slight was the resistance which it offered. The police streamed in. A four-wheeled cab was drawn up against the side; two horses were in their stalls; they seemed to be its only living occupants.

"Perhaps he's up in the loft."

The speaker pointed to a square hole which was in the ceiling. The chief called to the men who had gone upstairs,—

"Is there a door up there leading into the loft?"

"No, there's no sign of any loft up here. These are living rooms; there's no door which we can't open."

The chief turned to Mr James.

"How do you get up into the loft?"

"Jump up, always. I'm the champion jumper, I am; gives a bound, flies through the 'ole, and there you are. It's easy, when you've done it."

"Perhaps he's pulled the ladder up after him."

The suggestion came from one of the men.

"Go and fetch a ladder anyhow."

After a short delay one was brought; possibly obtained from a gentleman who was not wholly without a touch of sympathy for the troubles and trials of the minions of the law. It was placed in position. Without hesitation a constable ascended. Three or four lanterns from below were flashed on to the patch of darkness overhead, revealing the open woodwork of the roof above. The constable, having reached the opening, turned his lantern to the mysteries within.

"Can you see him?"

"No; if he's anywhere he's lying low, or else he's at the back."

Even as he spoke, a truss of hay, coming through
the opening, knocked him off the ladder on to the
ground.   The chief's voice rose in tones of command.

"Now, you chaps, up you go, and mind this time
you get him."

Heedless of the disaster which had overtaken their
colleague, three or four constables started to scamper
up the ladder.   Another truss of hay met the leader
at the top, but this time he was prepared.   In spite
of it he forced his way into the loft.   His companions
followed.   There were the sounds of a lively scrimmage overhead.   The chief went up to keep an eye
on what was taking place.   He found that already
the capture was effected.   Pollie was on her back;
two constables were holding down her legs, another
was at her shoulders, a fourth was adjusting the
handcuffs on her wrists.   He issued his instructions
to those below.

"Put a horse into that cab down there; we'll take
him in that."

"The door's locked, you know, sir; we can't get
it out."

"Then go and get one from next door or somewhere, only mind you're quick."

A cab was procured.   The prisoner, now sufficiently
quiet, was brought down the ladder.

By this time the news of the capture had spread.
In spite of the efforts of the police the mews was
thronged.   An excited crowd pressed about the
vehicle.   So soon as the prisoner appeared a voice
rang out above the surrounding discords.

"Why, it's Pollie Hills!"

The speaker was the lad Bob.   With a sudden mad
dash he had broken through the line of struggling
constables into the little open space in which the
four-wheeled cab was standing.   A policeman seized
him by the shoulder.   But even that unfriendly grip

was not sufficient to silence him. He continued to
shout, in spite of the desperate efforts he was making
to tear himself free.

"Why, you silly fools, that's not the man you're
after! What do you think you're doing? That's
Pollie Hills—that's a girl!"

An inspector, who had recently arrived, came forward out of the shadow. He peered shrewdly into
the prisoner's face.

"He's right—it is a girl; and what's more, it's the
girl who laid the information. What's the meaning
of this?"

"I tell you what it means," screamed Pollie. "It
means that he's a silly fool who wants to wring his
own neck; and so he can, and so I'll let him know!"
She pointed at the lad with her manacled wrists.
"That's him who gave me the ring and the bracelet,
and murdered the bloke in Embankment Chambers!
I've been dressing up to make out that I was him to
save his silly life, and now that, after all I've done for
him, he's gone and given the game away, I'm off.
You can take him and hang him out of hand!"

The inspector pressed the tips of his fingers against
Bob's chest.

"Do you mean that this is the man who gave you
the ring and bracelet which you brought to me?"

"Of course I do—what else should I mean? His
name's Bob Foster, and he gave me the ring and the
bracelet, and told me to keep 'em dark, but so soon
as I found out where he'd got 'em from I brought 'em
straight to you. But I'd have saved his life if he
hadn't been such a cuckoo as to give me away, and
himself too; and now, whatever comes to him he's
brought upon himself. Yes, Bob, you know you
have! Bob! Bob! after all I've done for you!"

She burst into a storm of weeping.

Before he had realised what was the actual purport

of her burst of invective, Bob found that he had a pair of handcuffs on his wrists. He stood looking at them like a man in a dream; then he glanced at Pollie, as if he expected the nightmare to pass away.

But the nightmare stayed.

It seemed, from the girl's tumultuous grief, that her heart was breaking.

CHAPTER VII

IN THE BOX

"THIS is the second time."

Recollection returned. He remembered where he was. He was in a box; that was the meaning of the narrow walls, the closely-overhanging roof, the darkness, the sense of oppression. Pollie had lifted him right off his feet, had laid him down within the chest. He recalled so much, and the grotesque consciousness of helplessness which had come to him on finding himself being lifted like a child, and that was all. The rest was vacuum. He repeated to himself, grimly, his own words,—

"This is the second time."

How long had he been where he was? What had happened since? All seemed silent. He thought of the journey along the roof, of the entry through the casement, of the appearance of the room as revealed by the flickering matches. Was he alone in it? Where was Pollie? What had taken place? He felt cold, cramped, worn, hungry, ill — perhaps hungry most of all. A faint light gleamed through the holes in the side of the box. Was it morning, then? If so, what had transpired during the hours which had intervened?

He tried to collect his thoughts, to arrange events in their proper sequence. He had a hazy recollection

of the girl telling him that he would have to make his
own way out. How, then, did she propose that he
should do it? He came into contact with something,
with several things, as he slightly shifted his position.
There was a hammer, a mallet, a sharp-edged instrument which he took to be a chisel, and a saw.

So she had put tools into the box with the idea of
his using them to further his escape. He had some
vague fancy of her having said to him something of
the kind. The idea was chimerical. He never had
shown to advantage when he had a tool to handle.
He was one of those men to whom has been denied
the capacity of becoming even moderately skilled in
any sort of handicraft. From the craftsman's point
he was a hopeless subject. He told himself that,
situated as he was, to expect him to win his way to
freedom by the aid of a hammer and a chisel was
sheer absurdity. He had scarcely room to move,
barely strength enough to raise the heavy hammer.
It was altogether out of his power to put it to an
effectual use. He might avail himself of it to attract
attention by striking it against the sides of the box.
But that would be to run the risk of passing from the
frying-pan into the fire. Who might not hear?
What undesired auditor, who might be keeping an
inquiring eye upon his whereabouts? And, in any
case, how should he explain his being where he was?

As he lay and pondered and smiled, even in that
very unamusing situation, there came back to him
other of the occurrences which had followed on each
other's heels in such quick succession. Mr Baynes'
visit—the tidings which he had brought of the
wealth which had been bequeathed to him by so
unexpected a testator. He smiled still more, and
shuddered. If there are such things as ghosts, and
the dead man's spirit was observing him in his present
situation, supposing such visitants to be capable of

sentient emotion, what, at that moment, must the "returned traveller's" feelings be?

A man of fortune, not to speak of family, shut up, a helpless prisoner, in a wooden box, like the youthful bride in that absurd old doggerel, "The Mistletoe Bough!" The thing was too ridiculous for serious consideration. Yet, what on earth was he to do?

He must do something—must make some effort to win himself out. To be found in there, like a rat in a trap, would be the crowning humiliation, and to stay in there and not be found, might be much worse. Picking up the chisel, making a half turn, he resolved to, at any rate, test the tool's calibre against the side of the chest.

As he moved, something so startled him that he dropped the chisel from his hand. Was it possible that some other living thing was shut up with him inside the chest? In moving, he had extended his right leg; the sole of his boot had inadvertently pressed against something, something soft and yielding; something which, unless he erred, returned the pressure. The contact filled him with such a degree of horror that for some moments he was bereft of the power of motion. What could it be? A delusion? If not, then what?

He must have been mistaken. There could be nothing there. Yet, though he assured himself that this was the case, it was odd how reluctant he was to put this assurance to the proof. Instead of an inclination to reach out with his foot again, his inclination was all the other way—to draw it as far up as the dimensions of his prison would allow. On the other hand, the suspense was hideous. On such a point it was impossible that he could remain in doubt, and exert himself to gain egress from the box. Even supposing that there was something, it was probably a mouse or a rat, or perhaps a rabbit. It was true that it had not felt like either of the three, but what else could it be?

After an interval, the duration of which was only too perceptible, he ventured out with his foot again gingerly. He could feel nothing. He put out his foot a little farther. What was that? He could feel something which, as it seemed to him, was possessed of a peculiar quality, though wherein lay its peculiarity he would have been unable to say. But this time it certainly did not return his pressure. His alarm had been foolish. It was probably something else which Pollie had put in with him; what it was he had forgotten, if he had ever known.

Again he took up the chisel, and once more applied its edge to the side of the box. It was frightful how weak he was. The silly shock he had received had left his nerves in a state of uncomfortable tension. He was trembling all over. Such was his exhaustion that he could put forth no strength at all, not enough even to make a feint of driving the tool's cutting edge into the wood.

And the whole time his thoughts were at the other end of the box—he did not wish that it should be so, but it was.

Presently something else occurred to him,—that there was a peculiar odour in the chest. It spoke volumes as to his condition to say that he had not noticed it before. His sense of smell was almost un-naturally keen. That he had been unable, on occasion, to escape from disagreeable perfumes had been for him one of the most disagreeable characteristics of the vagrant life which he had led. And there was a peculiar odour; of that there could be no doubt what-ever. Now that it had become perceptible, it took his nostrils by storm. It seemed to him that in it there was something ominous, alien, repulsive. In it, too, there was something familiar, an association with an experience which had, to say the least, been intensely disagreeable.

Was it possible that the thing at the bottom of the box had been alive and was dead, in a state of partial decay?

With a sort of gruesome curiosity, he put out his foot to feel again. What could the object be? It seemed to be of a rounded form—and distinctly soft. The more he pressed, the more it yielded.

It moved! And he moved too, withdrawing his feet so far as he was able, with a kind of spasmodic instinct. This time there was no mistake; the thing had been certain, the movement definite, distinct, though even now he felt that about it there had been something unusual, odd. What could the creature be? As, with whirling brain and palsied limbs, he asked himself the question, he was conscious that his panic fear was scarcely dignified; and with such slight force as still was his he did his best to play, at least in some degree, the man.

He might have succeeded had he had time. He was not in a condition to recover from a shock upon the instant; and when the shock became a continuous, an increasing one, recovery, even of the most partial kind, was altogether out of the question. The few remaining dregs of his manhood oozed clean out; there was nothing left to him but his shivering carcase.

The creature at the other end of the box was roused at last; and, being roused, was apparently unwilling any longer to remain dormant. It kept on moving, though its companion in captivity could neither feel nor hear it; indeed, the perfect stillness of its movements was not its least unpleasant feature.

It touched his boot; it was advancing from its end of the box to his. It not only touched his boot, it climbed on it with a curious gliding motion, which held him paralysed with horror. It advanced along his trousered leg, and yet he felt it still upon his boot.

He knew what it was now : it was a snake ! And, with the knowledge, he broke into shriek after shriek, and writhed and twisted, almost after the fashion of the reptile at his feet, as if he sought, by his sheer contortions, to get out of the box. But, after all, his power to shriek, like the rest of his powers, was small, and soon he was still, perforce. He lay shaking as with ague, as pitiable an object, if Howard Shapcott's ghost was there to act as witness, as that gentleman's eyes had ever lighted on.

The snake, as if startled by its companion's cries, continued for the moment still, but, so soon as the man ceased, it resumed its movements. Along the man's body it glided, with, seemingly, complete indifference to the fact that it was a man's body. The sensation was indescribable ; he never forgot it ; it was always with him, afterwards, at the back of his memory, needing but some casual touch to bring it to the front. The thing seemed endless ; its weight was no unconsidered trifle ; it was evidently a monster of its kind. Its head reached his ; he felt its breath upon his face ; something brushed against his cheek, softly, to and fro, with amazing delicacy of touch ; it was the creature's tongue. He knew enough to be aware, even in his state of partial brain paralysis, that the popular notion that a serpent's poison is secreted in its tongue is born of ignorance ; but at any moment the tongue might be exchanged for the poison fangs, those finely-pointed instruments which prick, at times, so slightly that a man need not be conscious of their contact until death's upon him.

Whether or not the creature, then, did more than salute him with its fondling tongue, he could not tell. Presently he ceased to feel its breath upon his cheeks. It continued its progress, passing down over his different features, and arranging itself, slowly, methodically, in a series of apparently innumerable coils between his face and the holes in the front of the chest. It had placed itself directly in his line of vision, so that it obscured the greater part of the small amount of light which found its way into the box, and there it remained quiet.

For some time he continued in a state of physical and mental stupor; it was only after a considerable interval that, venturing to open his eyes, he perceived that the creature had darkened the light. He realised its propinquity with feelings so intense as to be almost equivalent to a total absence of sensation. By degrees questions began to shape themselves vaguely in his brain. Since it proposed to remain where it was, what was its purpose? Was it watching him, playing with him like a cat with a mouse, waiting for the moment which suited it best, to strike—and to strike home? Or had it relapsed into torpor? He imagined that he could see its eyes, wide open, unblinking, glinting at him through the darkness.

His right leg was cramped. In the first sharp twinge there was an involuntary movement of the muscles. The reptile seemed to pay no heed. As the pain increased, he made a more vigorous effort to obtain relief. Still it evinced no symptoms of disturbance. In his agony he shook it, extending the limb to its utmost length, for the muscles were twisted into knots. Yet his companion remained motionless.

While he was still being afflicted by his latest enemy he heard a sound without—the sound of heavy footsteps coming up the stairs. They came into the room. In a moment, in his delight at being within reach of human succour, careless of what the consequences of discovery might be, he raised his arm, and struck as hard as he could against the lid of the box.

A human voice replied to the signal.

"Hullo, Jennie! Is that you?"

"No, it's me!"

There was a pause, perhaps expressive of the new-comer's astonishment. Then the same voice again, only drier.

"Oh, it's you, is it? And who are you? And what the devil do you mean by being in my chest?"

"Open the lid; let me out."

"You let yourself in; I suppose you can let yourself out."

"I can't. It's locked."

"Locked, is it? Then where's the key?"

"I don't know."

The newcomer had evidently approached the box, and, judging from his next remark, was apparently realising, with mingled feelings, the absence of the key. He tried the lid once or twice, to satisfy himself that it was actually locked.

"Well, this is a queer caper. It is locked; then who locked it? You couldn't hardly have done it from the inside, and what were you up to to let anyone else do it when you were inside? And where's the key?"

"I tell you again that I don't know. For heaven's sake, do let me out!"

"There you are, piling on the coal! Go soft! Let's take things in their proper turns. First of all, where's Jennie? If you've hurt her by so much as a pin-prick, I'll leave you inside until your bones are bleached."

"Do you mean the snake?"

"Of course I mean the snake; and you're right in using the definite article, the snake. She's the finest snake this world has seen. And where's my beautiful?"

"The snake is here."

"Have you laid on her a hurtful hand?"

"God forbid!"

"Right again—God forbid. I'd have hurt you if you had. So it seems that you're in luck. It's something to have a companion, even when you're shut up in a box—I speak from my own personal experience; and when it's a lovely creature like my Jennie, why, that's something like, that is."

The Gentleman was silent. He felt that the speaker had a point of view of his own; he was beginning to wonder if he had been sent to complete the process of driving him mad.

"Now that the first and most important point is in order, we'll pass on to the next. What about opening the chest? How am I going to let you out without the key?"

"Force the lid."

"Yes, that's all very well for you, but how about me? Forcing the lid won't improve the property. Who's going to pay for the damages?"

"I have money."

"Oh, you have got money, have you? It's a nice thing to have, is money; about as comfortable a thing, all things considered, as you very well can have. If we had more of it, Jennie and me, we'd be as happy as the days are long, as the saying is—ah, and happier. How much have you got?"

"I had four pounds."

"Had? I had a thousand pounds, and I've had it more than once; but, at present, it's behind—that and more. The question I put to you was not how much you had, but how much you've got just now."

"I believe I have the four pounds upon me now."

"Believe? Belief's all right; I've seen a lot of belief in my time; and I don't say a word against anyone believing anything. But I like to know. Don't you know if you've got that four pounds?"

"I had it on me last night, I've every reason to suppose it's on me now; but I'm tortured by cramp —half-paralysed—I can't feel to make sure."

"Then am I to understand that any damage I may do my chest by forcing open the lid you are willing and ready to make good?"

"Every sou! Only, for God's sake, make haste. If you knew the agony I've endured, and am enduring, you'd take pity."

"Agony! Is that a slur on Jennie? Or on my chest? You've had a free night's lodgings, and the best of company; I don't know what else you want."

The Gentleman groaned. Was the man a madman? Or had he been sent to torment him before his time? Heavy footsteps crossed the floor.

"Someone's been taking French leave with my tools. There's a cutting chisel, a hammer, a mallet, a saw, all gone; they weren't gone last night."

"Isn't there anything else you can use?"

"No doubt. I might use the lid of a teapot; or a toothpick. But I'm not proposing to use either."

"Make haste, man, do!"

"What's the hurry. If, as you say, you've been all night in agony, you must have got used to it by now. What difference does a few minutes more or less make? I've been in agony myself, ah, and for more than one night! I speak from personal experience."

"Are you a fiend?"

"May be. Don't know. A man never does know till he's been cut up for the doctors to examine; what he is inside's a mystery. Here's a cold chisel. Shall I use that?"

"Yes."

"Seems to me that you'd say 'yes' to anything. If I was to ask if I should use the stem of my best pipe you'd say 'yes' to that. You don't seem to have no sense, no proper reasoning. However, things being as they are, I don't mind making a trial with the cold chisel, just to oblige you."

The footsteps returned towards the box.

"Now, I understand that for any damage I do to this property—and it's valuable property, mind you, so don't you forget it—I understand, I say, that for any damage I do to this property, on your behalf, and for the purpose of letting you out of it, you'll see me paid."

"Yes."

"That is the understanding, between gentlemen as gentlemen?"

"It is."

"Very well, then. That's all right, so far as it goes. When I've put a bit of baccy into my pipe, and got a light to it, I'll set about the job."

The Gentleman waited, in hideous expectation. He seemed to be an actual witness of the production of the tobacco and the pipe; of the pipe's loading; he heard the scratching of the match; he saw the flash; smelt the perfume of the strong tobacco.

Then there came the voice again.

"That's it; that's something like, that is, the right article. Now I've got the taste of tobacco in my mouth I'm game for anything. Look out in there. I hope soon to have the pleasure of making your acquaintance, sir; the job is being started."

In less than half a minute the chest was open.

F

CHAPTER VIII

THE TWO MEN AND JENNIE

THE Gentleman, peering up, saw a hairy face looking down at him—a face of which the hirsute adornments indeed were so prominent a feature that, without the least assistance from Art, he might have taken a first prize in any exhibition of hairy men. Beard, whiskers, moustaches, eyebrows, the matted shock which covered his cranium, were all of extravagant dimensions. A pair of shining brown eyes glanced out from the centre of a thatch. It was only when his lips were open that you perceived he had a mouth.

"Glad to have the pleasure of making your acquaintance, though the meeting's a trifle unexpected. A man doesn't look to find a friend in his chest, not every day. Well, now the lid is open, are you going to keep where you are?"

The Gentleman tried to raise himself, and failed. He closed his eyes in a spasm of pain.

"I'm afraid I can't—get up."

"Can't, can't you? Does that mean you're expecting me to lift you? That's not in the bargain. Extra work, extra pay."

The stranger, passing his arm under the other's prostrate figure, lifted him out as easily as if he had been a child. He placed him on a wooden chair—on which he immediately assumed the attitude of a limp, lay figure. His head fell forward on to his chest, his legs curled fatuously in front of him, his arms dangled loosely at his sides. The stranger commented on this singularity in his appearance.

"A nice chap you look—well set-up, poker down your back, nose in the air; never was a chap what carried himself stiffer; and as happy as the day is long. Here's your hat."

He took out of the box his visitor's curiously-shaped top hat. It had not been improved by the obvious fact that the snake's course had lain right over it. Without, however, paying any attention to such trifling details as that it was bulged in at the sides and top, and that the nap had been copiously reversed, he planted it on its owner's head, a good deal to the back, and a little to one side.

"There you are, there's a tableau for you! When I was a young 'un, if it had been near the fifth of November, of immortal memory, and I'd been on the look-out for a guy, I'd never have let slip a chance like you. On the top of a tar barrel you'd have made a picture!"

The speaker returned to the chest.

"Hollo, here are those four sovereigns you spoke about. One—two—three—four golden shiners, left lying, careless like, at the bottom of my box. That's not business. I should have thought you'd have known better. Now I'll put 'em on the shelf here, in full sight of both of us, so that when I come to make up my little bill, and you come to settle it, we shall both of us know there's been no cheating."

Suiting the action to the word, he once more went back to the chest. Stooping over it, he addressed himself to its remaining inmate.

"There you are, my beauty, looking, like you always do, a dream of loveliness. Do you feel your-self slighted because I've left you to the last? Never mind, my Jennie, it isn't every day you get a lantern-jawed living skeleton in there along with you, and you've got sense enough to know that in a procession, on or off the stage, the stars come last. He's a super, that's all he is, so that's how I come to give him my attention first; but now the stars are on. Come along, my sweetheart."

The snake, perhaps trained to obey his voice or his touch, reared its head out of the chest, and, gliding round the speaker's neck, began to wind itself in coils about his body. The man rose, with the creature girt about him, and, from a jug which stood upon a shelf, he filled a bowl to the brim with milk. Sitting on a stool, holding the bowl in front of him, the snake consumed its contents while its owner continued talking.

"I'm a snake-charmer, among a few other things. At least, I train and educate snakes for snake-charmers, which is the same thing with a difference. Me and Jennie, we do the work between us, don't we, Jennie? That's right, my girl, sup it up. That's genuine milk, that is, not pumped on chalk dust. I bring it in with me every morning—a pint from a special cow. She likes it as a pick-me-up; she'd miss it as much as I'd miss my pipe. Talk about snakes having no sense and dogs having it all, we can tell you different from that, can't we, Jennie? I've seen something about snakes, and I've seen something about dogs; and I tell you that for sense a dog's not in it with a snake. As for Jennie, she's one in a million—she's got more sense than all the dogs that's living. Her pupils, and mine, have performed before all the crowned heads of Europe—and made 'em gape, some of 'em, I give you my word. There isn't a snake, from the worst-tempered to the nicest-mannered, that she—and me along with her—wouldn't make a perfect scholar of, and in less than no time, so to speak. She's given a complete education to a few hundreds, so I ought to know. Haven't you, old girl?"

He stroked the creature's head with his huge hand, as caressingly as a mother might have stroked her child's.

"There's sometimes as many as fifteen or twenty in that living box along with her, all sorts and sizes and tempers. But she keeps 'em in order; she's got the knack of it, she has. Only once has two of 'em started eating each other while she's been round, and then, when she found out what was going on, she pulled the one clean out of t'other. Straight! I seen her done it."

He glared at his auditor as if defying contradiction. Nothing could have been further from the Gentleman than any intention of the kind.

"Though, mind you, it's perhaps as well for you that Jennie was the only companion you had last night. Grass snakes, and puff adders, and such, aren't nice bed-fellows, even when their fangs are drawn. They've got a natural antipathy to humans, and, when they've got a chance, they like to show it. But, as it happens, I sold the last snake I had last week to Loolah, the Queen of the Serpents. A nag, it was—cobra, you'd call it; hardly under five foot. It killed the chap who drew its fangs, so, of course, that makes it extra valuable—it makes a good line on the bills. A friend of mine's bringing me over a fresh lot, in the *Senegambia*, due next week. An assorted lot, I told him to bring; but, until they come, I'm out of stock; which, perhaps, was just as well for you. Finished, have you, Jenny? That's right, my beauty. You make yourself comfeys, and have a snooze!"

The serpent, withdrawing its head from the now empty bowl, seemed, as he said, to be composing itself to sleep. Fondling its huge coils with both his hands, he gazed down at it with a mixture of pride and affection, which, in one of his appearance, and as displayed towards such an object, was sufficiently grotesque.

"Talk about affection; she's more affectionate than any wife I ever heard of. She's never done me a bad turn, nor spoken to me a hard word all the time I've known her. They say women are snakes. Well, all I can say is that I'm glad snakes aren't women. I'd be in a pretty bad hole if they were. Jenny here's worth more than a street full of women—yes, and that by a good deal. She don't cost much to keep, and nothing to dress. She don't go gadding about, always in search of excitement. She don't spend your wages for you; and she don't make a noise, not like a dog or a woman would, disturbing the neighbourhood. Let them as want a wife get one, and then go and hang themselves, after they've cut her throat; which, very soon after they've got her, is what they'd like to do, if they don't always do it. No woman need apply to this address; give me Jennie."

He continued smoking, stroking the monstrous reptile in silence, as if ruminating on its manifold virtues. Presently he changed the theme.

"How are you feeling now?—less of a guy?"

"Thank you, I am a little better; I am beginning to feel again that my limbs are, to a certain extent, my own. Such a night as I've spent in that chest of yours, is—not restful."

"And yet there are worse beds. I've had 'em—and you may have before you've done."

He puffed his pipe and eyed his visitor.

"How did you come into that chest?"

The Gentleman smiled, with a touch of his characteristic whimsicality.

"Through the lid."

"Ah! That way! I see!"

There was more puffing at his pipe, punctuated by the use of one of his fingers as a tobacco stopper.

"I'm told that there was quite a little party here last night—a surprise party; and if I hadn't been told I'd have known. The window was closed when I went, likewise the door. They talk about a woman's neat ways—when a man's got neat ways a woman's neat ways aren't in it beside of his. I have got neat ways; I like to keep my place just so; this ain't my notion of a tidy place, not by no manner of means." He glanced round the room. "It seems to me as if somebody had been trying all he knew to turn things upside down."

He relapsed into silence; the Gentleman volunteered no observation, so the smoker asked another question.

"Why did you get into the chest?"

"To keep out of the cold."

"To keep out of the cold? I see. It was that way." Another interval. "I have been told that those who were here were mostly police. You don't happen to know what they wanted?"

"My good sir! Do you credit me with a fund of information?"

"You are feeling better, I can see. I'm also beginning to get a notion of the kind of man you are. You're a man of education—a scholar, like the snakes we train—Jennie and me." Another pause. "There was some mention of a murder." Silence.

"Have you heard anything about a murder?"

"There's murder in the air."

"Now? Do you mean the murder of me, or of you? You don't look like murdering me; but looks are deceiving. I've known little chaps, whippersnappers, what looked as if they hadn't got strength enough to kill a fly, do for men three times their size—giants. Of all the gambles I know, there's few to beat the game of killing. When that game's on, you be careful before you back a man to win."

There was a pause of considerable duration. The man with the serpent wound about him regarded the other with continuous, unblinking gaze, as if he found him a study of quite unusual interest. The Gentleman endured the scrutiny with a serenity which, in itself, was curious. So far as limpness went, his resemblance to a lay figure had become distinctly less, but he still presented a pitiable appearance. His features were pinched and blue with cold, tremulous with weakness, drawn with hunger; it was obvious that he still endured extreme discomfort.

The hairy man, taking his pipe from his mouth, pointed at him with the stem, as if he sought to drive, with its aid, his remarks well home.

"The way in which I look at it is this—correct me if I'm wrong; I'm only throwing out, as it were, a line. I find a party in my chest, locked up in it, along with Jennie here. I'm informed that the police have been here inquiring for a party. When I ask myself what connection the party in my chest has with the party the police have been inquiring for, I have to find an answer. What do you think?"

"I have no thoughts."

"No? So I see. That's the kind you are—a man of education. And if education don't teach a man when and how to shut his jaw and keep it shut, what it does teach him I'd like to know."

The speaker, smoothing with one hand the coiled serpent, waved, in the other, his pipe in the air in a vague, impersonal fashion, as if to denote that, in what he was about to remark, he was merely advancing a general and self-evident proposition.

"Now, as to murder, as murder, I don't think nothing—nothing at all. I've seen too much. As Lord Alfred Tennyson says, 'It's all in the name.' There are parts of the world, plenty of 'em, where it's looked upon as one of those little accidents which dot man's pathway to the silent grave. Why, I've been where it's been considered good manners to kill a man what wouldn't drink with you; yes, and where not to have killed a man would have been a black mark against your character. It's a question of climate—that's what it is—climate, all climate; and I've been in all sorts of climates, so my outlook's wide. I don't ask you what kind of accident you've had, or even if you've had an accident at all—I ask you nothing. Because I take it for granted that with you, like with me, and like with all the rest of us, there are subjects on which you'd rather be asked no questions; so I ask none. But what I'm coming to, and what I've been coming to all the time, is that this is a case which has a personal application. There's been damage done—to property, my property, Now, I'm not in a position to see my property damaged, and say thank you and nothing more. What I've got to do is to know where I'm to look for that damage to be made good to me."

" You're to look to me."

The Gentleman tapped himself on the chest with his quivering fingers.

" And that's where I am looking, so now you understand me; and my respect for you is growing because you do understand me. A fool I can't abide; I'd sooner, any day, have dealings with a rogue. That you're no fool is plain."

" You do me honour."

" Not more than you deserve. Now what would you say is the amount of damage that's been done ? "

"That is a point on which you alone are in a position to speak with authority."

"Well that's what I was thinking; strange how you and me do understand each other. Casting my eye right round the room, looking at this, that, and the other, knowing what I value 'em at, and what damage has been done to them, I should say that the amount was—"

"Yes?"

There had ensued a portentous pause; the hairy man had fixed his eyes upon the ceiling, as if he sought for inspiration there.

"At the lowest figure, mind you, at the very lowest possible figure, I should say that the amount was five pounds nine and four?"

"An extremely moderate estimate."

"That, again, is what I think. I never met a man for whom I felt so much respect on such a short acquaintance. The question is, where's that five pounds nine and four to come from?"

"I will undertake to let you have it in the course of to-day."

The hairy man returned his pipe to his mouth with somewhat of an air of dubiety.

"There are four sovereigns of yours now upon that shelf."

"There are?"

"Haven't you got any more coin on you?"

"I believe I have twopence in the lining of my hat. I had."

Judging from the expression of so much of his countenance as was visible, the hairy man was reflecting.

"When I meet a man who understands me, like you do, I'm generous—to a fault! That's my character. So what I say to you is this—you give me those four sovereigns, and I'll let you off the one pound nine and four, without a sigh."

"And I may keep the twopence which, I believe, is in the lining of my hat?"

"You may, and I won't murmur."

"You're too good."

"Don't mention it—not a word. That's me!"

The Gentleman rose to his feet painfully.

"And do I understand that I'm at liberty to go?"

The other also rose.

"You are—at perfect liberty! As free as the air! To go just exactly where you please."

"You are again too good."

"I'm glad to have had the pleasure of making your acquaintance."

"That is a gladness which we share."

"I'll say no more." He put his hand up to his mouth. "I hope the slops won't get you."

"That hope is something else we share. And I, on my part, thank you from the bottom of my soul for your splendid hospitality."

"Now don't — you're welcome. I begrudge you nothing you've had."

"And I wish you, sir, good-day."

The Gentleman removed his battered hat with as courtly an inclination as, at the moment, was at his command, then moved stiffly towards the door. The other checked him.

"Stop! Just one moment! If you were to come across that one pound nine and four, and were to feel that I haven't had all I ought to have—why—"

He stopped at the psychological moment.

"Precisely. You may rest assured. It is altogether impossible that I could ever feel that you had had all you ought to have. Again I wish you, sir, good-day."

"Won't you say good-day to Jennie?"

"I wish also a good-day to—Jennie!"

Saluting the serpent, which, by an odd coincidence, at that moment advanced its head from its post of vantage under its master's chin, putting out its feathery tongue, and opening its eyes, as if to proffer him a farewell greeting, the Gentleman passed from the room, walking a little shakily.

CHAPTER IX

THE HEIR AND THE LAWYER

THE biting, blustering frost of the night had been exchanged for a thaw which was almost colder. The heavy masses of snow which had fallen were being transformed into pools of slush. The skies were leaden. A suspicion of sleet embittered the air. As the Gentleman stood in the doorway, looking up and down the lane, the dreariness of the scene chilled him afresh, to the very marrow in his bones.

No one had perceived him in his journey down the rickety stairs. No one was in sight in the miserable lane. Going out, after a moment's hesitation, into the cold and the wet, he turned his steps towards the thoroughfare at the end. A woebegone appearance he presented. His boots were leaky; his patched trousers fluttered draughtily about his legs; his threadbare frock coat was buttoned with suspicious closeness to his skin. His battered hat and chipped eye-glass, on its piece of twine, lent touches of incongruity to a costume as ill-adapted to a pedestrian excursion on such a day as it easily could have been.

His outward bearing was in harmony with his inmost feelings; life, at that moment, was at its lowest ebb.

"Shall I make an end of it?" he asked himself. "Shall it be the river? I can't be colder there than here."

But the thought passed as quickly as it came. Not yet. At least, not yet. There were other things which must first be done.

He smiled, grimly, wryly, as he thought of the lawyer's visit, and of the tidings he had brought. How much had transpired since then. In the passage of a night how markedly had bad become worse. He a man of fortune, with even his four pounds gone, and a possible twopence left in the lining of his hat! At the thought he laughed aloud. A hurrying woman, hearing the sound, glanced at him apprehensively, deciding that so miserable a looking object who could laugh must certainly be mad.

Baynes had been enjoying a jest, or even laying a trap. That Shapcott had left him his money was, on the face of it, absurd. But he had promised Baynes that he would call on him, and he would; that was one of the things which must be done.

The lawyer's chambers were in Lincoln's Inn Fields. It seemed a long trudge from that Southwark rookery. The Gentleman thought that he would never get there. More than once he had to lean against a wall or some friendly post. It was as much as he could do to resist the temptation to sit upon the sloppy pavement to rest. As he crossed the river he looked down upon the swirling waters with eyes which were not devoid of longing. A watchful constable, noting how he leaned against the parapet, moved him on; there was a suggestiveness about his attitude which did not inspire official confidence. By the time he reached his destination he felt more dead than alive; it was only with difficulty that he could place one foot before the other.

He looked about the Fields, holding on to the railings in front of a house, like a man in a dream. How well he had known it once upon a time! Had that knowledge only been part and parcel of a dream?

He had had chambers in the Fields; handsome chambers they were, furnished as became a man of artistic taste. There, on the other side, was the house. Was it possible that he had ever looked out upon the gardens through windows of his own? Was it another man who had done this, or was it he? He told himself bitterly that it was another man.

He stumbled along towards Mr Baynes' chambers. At the door he hesitated, conscious of a curious unwillingness to pass the portal. The idea of such a figure as he was intruding on the decorous premises of a lawyer of the old-fashioned type, who prided himself on having a connection exclusively among families of position, appealed, even then, to his sense of whimsical humour. It was this, as much as anything, which urged him to advance.

As he stood at the office door, a spruce young clerk, the sole occupant of the small ante-room, looked up at him with astonished eyes, as if amazed at the insolence of his intrusion. His manner was brusque.

"What do you want?"

"Mr Baynes."

The youth hesitated, regarding him more and more askance.

"What's your name?"

"No name."

"What's your business?"

"Mine."

"Mr Baynes won't see you if he doesn't know your name or your business."

"Tell Mr Baynes that the person who, last night, promised to come, has come."

The young man hesitated again, then vanished, presently reappearing with surprise on his countenance, mingled with something else.

"Step this way."

The Gentleman stepped that way, passing through a room in which were clerks who eyed him with looks in which curiosity was blended with amusement. He passed through a door which the clerk held open, Mr Baynes advancing briskly towards him the moment it was closed.

" Mr Blaise ! Is it you ? "

He stopped when he had come half-way, as if in stupefaction.

" Yes, Baynes, it is I—what's left."

" But—how come you in such a plight ? "

" Owing to a long chain of little accidents which I have no time to recount to you in detail."

" But—"

" ' But me no buts,' Baynes. Have you a chair ? "

" Come and sit in this one. I think you'll find it comfortable."

" Comfortable ! It's a dream of bliss." The visitor sank down on to the roomy arm-chair. As he allowed its capacious fastnesses to embrace his wearied frame he closed his eyes with an expression which almost approached to ecstasy. " I wonder how many centuries have slipped away since I sat in such a chair."

The lawyer was observing him with obvious concern.

" Mr Blaise, aren't you well ? "

" Cold, Baynes, cold."

" Draw closer to the fire."

" And hungry, Baynes, hungry. Have you ever noticed how, when the internal furnaces are insufficiently stoked, the general temperature goes down ? "

" Have you—not breakfasted ? "

" Nor supped, nor dined. I'm starving, Baynes."

The lawyer bustled to a sideboard.

"I've some cake and biscuits, and some wine and brandy, nothing more solid; but I can quickly send out and get something."

"The cake and biscuits will serve, and the wine and brandy; they will form a foundation on which to lay something solid later on."

Mr Baynes produced two plates—cake on one, biscuits on the other. He placed them on a small table, which he drew up to the visitor's side, The Gentleman, taking two handfuls of biscuits, began to devour them with ravenous haste.

"A glass of this port wine, Mr Blaise, will do you good."

Accepting the offered glass, the Gentleman, swallowing the contents at a draught, again closed his eyes, as if in ecstasy.

"Nectar, Baynes—just nectar. It goes through my veins like grateful fire. But one glass is enough for a time. A second will go to my head, and you'll think I'm drunk."

The lawyer was still observing him with doubtful eyes.

"Aren't you very wet?"

"To the skin."

"Mr Blaise! You must change at once into drier things."

"Whose? My entire wardrobe I am wearing. Don't you admire my tailor, Baynes? His style is a little archaic, but not without distinction. Why, man, it's nothing to be wet when you're inured. When you've tramped about in the rain all day, and spent the night in the same clothes, under a damp arch—when you've done that, say, a score of times, whether you're wet or dry becomes a factor hardly worth consideration."

The lawyer was standing before the fire. He blew his nose.

G

"I'm afraid you've had some uncomfortable experiences."

"I've had glimpses into hell."

"Mr Blaise, why have you kept yourself hidden all this time? Why did you not at least communicate with me?"

"Baynes!"

There was something in the intonation of the name which seemed to appeal eloquently to the lawyer's bump of understanding, and to supply him with all the answer which he needed. The Gentleman, having finished the biscuits, commenced upon the cake, eating now with not quite so uncomfortable a display of ravenous appetite. His companion stared at the fire. When he spoke again, his voice was softer.

"Well, Mr Blaise, thank God it's all over now."

"Be thankful, Baynes."

"And you, Mr Blaise, be thankful also." The Gentleman was still. The other went on. "You are still young, in the prime of your age; you have the best of your life in front of you."

"You think so?"

"Beyond a doubt. You are much younger than I am. I do not call myself an old man; I allow no one else to call me so, either. I hope to be in the full enjoyment of my powers for many years."

"You are different."

"Not a bit of it. I trust you have not cultivated a vein of morbidity, Mr Blaise."

"I have cultivated nothing. Do I look as if I had?"

There was a pause; when the lawyer continued, he held himself a little sturdily.

"You must pardon me if I speak plainly; but I have always known you to be a gentleman and a man of honour—of almost too nice honour, on certain points. I am convinced that, whatever you may have endured, you have done nothing unworthy of Blaise Polhurston."

"Ah!"

The monosyllable was pregnant with a weight of meaning. The lawyer, glancing quickly round towards the speaker, seemed to be endeavouring to gain a glimpse into all that it conveyed.

"You have had tempestuous passages."

"Precisely; I like the phrase."

"But it does not follow that a man is any the worse for having passed through cleansing fires."

"Are all fires necessarily cleansing?"

There was another spell of silence. Picking up the tongs, the lawyer placed a piece of coal in its proper position amidst the flames.

"They are awaiting, with great eagerness, your coming home."

"To my home?"

"Yes, to your home."

"Is it as it always was?"

"Yes, only you are missing."

"I should feel a curious hybrid there."

"You know that you are talking nonsense."

Another pause. The conversation consisted of pauses chiefly, as if each were employing them to seek out the smallest number of words with which to express the largest amount of meaning.

"Is my mother still alive?"

"Indeed, and hale and hearty. She has you always in her thoughts—and prayers."

"You appeal to my sentimental side. Or do you credit my memory with weakness?"

"I fear that your memory is overstrong. It is a thing greatly to be desired—the power of forgetting."

"Ah! Nothing is forgotten, Baynes—ever. The things which we do in our cradles rise to confront us as we are sinking into our graves."

The lawyer had seated himself in an arm-chair on the other side of the grate, and was rubbing his hands slowly, one against the other. He kept his eyes fixed upon the flames, as if he desired not to make his visitor uncomfortable by regarding him.

"Your mother has greatly changed. I never see her but she asks if I have news of you. She continually urges me to find you out. My failure to do so has given me pain, because there is something in her eyes which knocks at my heart. I am convinced that her ears are always attuned to catch the first sound of your returning footsteps; and that she lies awake at night because, in the darkness, she can see you clearer."

"There is in you a vein of sentiment—of imagination, Baynes."

"Do you gibe at me because I tell you that your mother suffers—that the only thing for which she lives and hopes is the hour of your return?"

"Do not allow yourself to become heated, and do not quarrel with me at our first re-meeting."

"God forbid!"

"When my mother bade me go, I warned her, in obeying her, that I should not be able to show equal obedience when she might bid me to return."

"Can you not forgive your own mother, Mr Blaise?"

"Look at me, and supply your own answer. My mother has made me what I am."

The two men were still; they both stared at the fire. Presently the lawyer spoke again.

"Your sister looks for your return with almost as much eagerness as your mother; she, at least, has never injured you."

"Is she still unmarried?"

"She's a widow."

"A widow? No!"

"Yes, with one child, a daughter, who is herself a promised wife."

"Ye powers! These little facts accentuate the passage of the years. And yet you tell me that I'm still young." This time the lawyer made no reply. "Well, Baynes, I'm a living exemplification of the fact that self-made men are sometimes failures."

"How comes it that I find you in such a plight?"

"On my word, I can hardly tell you. Although experience has taught me that my career is but a commonplace, but one of a mighty army, I fancy that were it placed between the two covers of a book there are quite a number of people who would take it for a romance. Don't tell me that there is no such thing as ill-luck; I know better. I've sought for fortune in all the quarters of the globe; everywhere I've failed to find it, because it was written that I shouldn't. It's now more than a year ago that I returned from Melbourne, with scarcely a shoe to my foot, and with five and sevenpence in my pocket. I've lived — which, being interpreted, of course, means starved — on little more than that sum, since. So, when last night you appeared upon the scene with such a tale, it was not strange that I felt—incredulous."

"The tale is true."

"That I'm a man of means?"

"That you're a man of means. Howard Shapcott has left you all he had."

"But, in the name of the Prophet, why? What's the meaning of this new display of Fortune's irony? He never loved me."

The lawyer was pressing the tips of his fingers together, apparently as an aid to meditation.

"Frankly, I don't believe he did."

"I'll swear he didn't."

"On the rare occasions on which he mentioned

your name in my hearing, it was always with—
resentment."

"I can believe it—easily."

"During recent years I have seen him but seldom.
Although I was his legal adviser, we were scarcely
congenial spirits. I have never forgotten how, in
the years gone by, he wronged you, and another;
on your account I have never ceased to resent what
he then did, nor have I endeavoured to conceal from
him what were my feelings."

"Thank you, Baynes."

"On more than one occasion I have requested him
to transfer his business to other hands, but he has
refused point blank. I think that he respected my
honesty and independence, being quite capable of
appreciating in others the qualities which he himself
neither possessed nor desired."

"Let us say nothing of the dead but what is good."

The lawyer shook his head, as if uncertain if it
were always possible to act up to the letter of the
admonition.

"My theory is that his conscience pricked him, and
that this devising to you of his possessions was intended to be a deed of atonement for the injury
which he was conscious he had done you."

"Not so. It was vengeance he proposed."

"Vengeance ?"

"It was his intention to strike me, out of the grave.
He had the eye of a seer. He foresaw that, to make
me the gainer by his death, would be to enable him
to keep on scoring, off me, to the very end."

"How so ?"

The Gentleman shrugged his shoulders.

"Quite possibly it's merely my own fantastic point
of view—I won't attempt an explanation. In plain
terms, Baynes, what in the material sense does Shapcott's action mean to me ?"

"First of all, I'll show you his will." The lawyer handed his visitor a document which he took out of a safe which had been let into the wall. "This, as you will see, is the attested copy; the original is in Somerset House. I am named therein as sole executor. When I gathered the tenour of the instrument, I thought it advisable, in your absence, to avoid the risk of possible complications, to prove it at the earliest possible moment; that has been done."

The Gentleman unfolded what had been given him.

"It's short enough."

"Yes; it's brevity, from the legal standpoint, is not its smallest peculiarity."

Blaise Polhurston set himself to read the last will and testament of the man, for the arrest of whose murderer a reward of two hundred pounds was being offered.

CHAPTER X

A PROBLEM IN MURDER

THE will, as Mr Baynes agreed, was certainly sufficiently brief; there was a complete absence of that curious tautology which is apt to be a characteristic of such a document. Nothing could have been clearer, more to the point.

"I give and bequeath all that I possess at the time of my death to Blaise Polhurston, son of Rhoda Polhurston, of Polhurston, in the parish of Trevennack, in the County of Cornwall; and I desire that my lawyer, Henry Baynes, shall act as sole executor of this my last will and testament.—JOHN HOWARD SHAPCOTT."

That was all, with the exception of the date, and the signatures of the two witnesses—George Mason and Everard Lucas. Not a lengthy screed, nor one whose purport it was hard to grasp; yet the man to whom it was of such capital importance read it over and over again, as if he strove to read between the lines a meaning which certainly was not upon the surface. Once, twice, thrice, four times he read it; then laid it on his knee, and stared into the fire; then read it again and again; then turned to his companion, who had watched in silence.

"Not a penny to anyone else?"

"Not a penny."

"It's very odd.   Is there no one who has claims?"

"That I cannot say.   'Claims' is a word to which can be attached a variety of meanings.   He has told me, more than once, that he had not a blood relation in the world."

"Can I do with it as I like?"

"Entirely ; it is yours without reservation."

"Who are the witnesses?"

"George Mason is a porter who, that night, happened to be on duty in the building.   Everard Lucas, of whom I have some knowledge, occupies the set of chambers which adjoins Shapcott's."

"Ah! A big man, with a black beard?"

"That is he.   You know him?"

"No.   I may have seen him.   Go on."

"Lucas has a clear recollection of the whole affair. Shapcott called him in as he was going out after dinner.   Mason was already there.   When Mason had signed, Shapcott gave him half a sovereign. After he had gone, Shapcott read the will to Lucas, declaring that Blaise Polhurston was a man to whom he owed a good deal, and to whom he thought it extremely possible that he should owe much more."

"He had a prophet's vision."

"A prophet's vision? I don't understand what you mean in this connection."

"It's only a figure of speech.   Go on."

"Lucas says that he does not know if he was in earnest in speaking of his debt to you.   He was in one of his sarcastic moods, when it was not easy to perceive what it was he really intended you to understand."

"I know."

"The original is a holograph.   He told Lucas that he had just written it."

Blaise Polhurston stared at the fire.   There was a

suspicion of a smile about the corners of his lips. He picked up the will, and read it again.

"Have you noticed the date? Less than a week before the end?"

"That is something in the nature of a coincidence. I have searched his papers, but can find nothing hinting at the existence of any previous will. It is fortunate for you that one was made when it was."

"As I say, he had a prophet's vision."

"He could hardly have foreseen that in less than a week he would be murdered."

"Who knows?"

"Do you suggest that he was being threatened, and was aware that he went in peril of his life?"

"I suggest nothing. I only wonder. The whole thing, Baynes, is so very odd. By the way, what's it worth? The estate, I mean, to which I'm such a fortunate successor."

"There are sound investments of the present value of about £80,000, which bring in an average return of five per cent."

"Then I'm the actual possessor of an income of £4000 a year?"

"About that—rather more than less."

"How extremely funny!"

"There are, also, certain personal effects, and a cash balance at the banker's of £2489, 16s. 3d."

"£2489, 16s. 3d.! Ye powers! Can I have that at once?"

"Certainly. If you like, I will go into details with you now, and hand you over everything. The estate is in order, and at your service."

"No, thanks. There will be time enough for that a little later." He stood up, fronting the fire, and tapped with the will against the mantelshelf. "I suppose there is no doubt that Shapcott was murdered?"

"Not the slightest. He was shot through the heart."

"Through the heart? That was how it was. Was his death a painful one?"

"According to the medical report he must have died in an instant. There was no time for feeling."

"Suicide was out of the question?"

"Entirely. The theory is that he was seated in his chair, and that, as he was turning, the murderer shot him from behind. The bullet passed downward through his lungs and heart, and out into the floor. It was found embedded in the floor. No man attempting suicide could give such direction to his shot."

"I suppose not. Is there any clue to the murderer?"

"So far as I know, none. The police talk, but that is all. I have offered a reward of £200."

"Oh, it is you who have offered the reward?"

"Yes, on your behalf."

"On my behalf!" Mr Polhurston spun round towards the lawyer like a teetotum to which has been given a sudden twirl. His attitude denoted unqualified surprise. "Do you mean that I—I!—am to pay £200 for the discovery and conviction of Shapcott's murderer?"

The lawyer seemed to consider his answer. It came from him a little stiffly.

"Of course, I quite understand that my action was unauthorised. You are perfectly at liberty to repudiate what I have done. I merely thought that, since you were the chief gainer by the man's tragic end, you would not be unwilling to stimulate the search for his murderer."

Blaise Polhurston stared at him as if he could not altogether make him out. Then he broke into curious laughter. Dropping back into the armchair, he threw up his arms.

"Baynes, unconsciously you're really funny. The

idea of my offering a reward for the hanging of Shapcott's murderer is, in its way, a stroke of humour."

"How so? It seems to me, with your permission, that, under the circumstances, to have withheld such an offer would have been scarcely decent."

"Exactly; you're quite right, Baynes. Don't be huffy, man. It's only my keen eye for a jest. I am perfectly willing to pay the man, or woman, who hangs the scoundrel, a couple of hundred pounds. He, or she, will have earned the money. Nay, I'll go further; whoever brings Shapcott's murderer to the gallows, I'm content to make my heir-at-law. I might as well."

"Your sense of humour has not grown less keen, Mr Blaise. For my part, I fail to see in such a subject the making of a jest. But then I'm dull."

Mr Polhurston was still. His legs were stretched straight out in front of him, his hands were on his knees, his head had fallen forward on to his chest; his eyes stared at the fire; on his face was the queerest smile. His manner, when he did speak, was a thought sententious.

"My dear Baynes, when a man has known the heights and depths, and all sorts of weather, he likes his humour mordant. The more drastic it is, the more it stings; and the more it stings, the sharper the relish. It is like some Anglo-Indians, with hob-nailed livers, who can only taste the strongest curries. But to return to our muttons, Baynes—uncurried. I suppose that the motive of this monstrous crime was common robbery?"

"That is not by any means so clear. If so, for some reason or other, possibly because he was afraid of interruption, the villain took to flight before his purpose was completed. There was a large quantity of portable property within easy reach, which he

might without difficulty have carried away and
readily converted into cash, which he left untouched.
Indeed, we have only been able to point clearly to
two pieces of property which are missing—a couple
of trinkets. One was Shapcott's ring—"

"It was his ring?"

"Yes; where did you see it mentioned?"

"Oh, on the bills."

"The scoundrel must have drawn it off his hand after
he was dead; he always wore it on his little finger."

"It might have fallen from his finger on to the floor,
and, from that position, caught the—scoundrel's eye."

"It might, but it's not likely. The other thing
which was taken was a bracelet; and about that
there's something curious. When first the discovery
was made, on the table in front of Shapcott was an
envelope which, the presumption is, he had been
examining at the very moment that the attack was
made on him. On it was an indorsement: 'In this
envelope there is a bracelet'—then it went on to
describe it with detailed minuteness, winding up with
the statement:—'It is one which I gave to Helen
Fowler.'"

"Baynes! What!"

Half-rising from his chair, Blaise Polhurston
turned upon the lawyer with a face which had
become all at once transfigured.

"So the endorsement ran:—'It is one which I
gave to Helen Fowler, and which was found upon
her wrist when her body was taken from the weir
pool.' As no trace of such a bracelet was anywhere
to be found, it has been concluded that the murderer
took it, together with the ring, doubtless unaware of
that minute description on the envelope, and that,
in it, therefore, he was taking something, which,
quite possibly, may be the means of bringing him to
the scaffold."

"And that was Helen's bracelet!—and she had it on her when they found her dead! What a nightmare!"

Mr Polhurston sank back into the arm-chair, all the life seeming to have suddenly gone out of him. Mr Baynes, with an air of the deepest concern, came and touched him on the shoulder, as if to rouse him from the lethargy into which he appeared to have sunk.

"Mr Blaise, I am so sorry. If I had had a notion that after all these years your feelings on the subject were still so keen I'd have spared you the allusion."

"Helen's! Helen's bracelet! I believe that I remember it. How she flashed it at me, and how, when I asked who had given it her, she laughed in my face."

"Mr Blaise!"

"It all comes back to me. How she twisted it about her wrist until I could not help but see it. When it caught my eye I was all hot in a moment. 'Who gave it you?' I cried. 'A friend,' she said. 'What friend?' 'Ah, that's tellings.' And she wouldn't tell, although I persisted in inquiring. So I lost my temper, and left her in a rage. And Shapcott was the donor? If I'd known it I'd have killed him—then."

"Mr Blaise!"

"Yet there are those who say there is no God; why, there's a time in a man's life in which he knows that it is God's finger which has touched him." He glanced up at the lawyer with lack-lustre eyes. "But come, Baynes, you look so serious I fear I also must seem grave. We'll none of that; at this moment all the world's so full of promise." He threw back his head, as if resolved to put away from him the burden of discomforting memories. "Now tell me all about it.   Did no one see this—bloodthirsty ruffian ? "

" No one, as it seems."

" Nor hear anything ?   For instance, the shot that did the deed ? "

" Not a creature, so far as can be ascertained."

" Was it fired from an air-gun ? "

" An air-gun! No! From a revolver. And a sufficiently formidable example of its class. It was found about three feet from Shapcott's chair; as if the assassin, suddenly alarmed, had let it fall and had been too panic-stricken to stay to pick it up. It's a big, murderous-looking weapon. On the stock are initials, which are, quite possibly, the murderer's; and that's another clue which may serve to tighten the noose about his neck."

Mr Polhurston looked round at the speaker.

" Initials ?   On the revolver ?   Where ? "

" On the stock. The weapon is an old one; at least it shows signs of having seen a considerable amount of service, but the initials are comparatively fresh, as if it had recently changed hands, and as if the new owner, in his first pride in his new possession, had desired to announce it to all and sundry. If so, it is quite within the range of possibility that his indulgence in that desire will hang him."

" May not what you call initials be merely a tradesman's mark ? "

" Scarcely. For one thing, they are so large as to almost amount to a disfigurement. A man would not buy a weapon on which somebody else's initials were so ostentatiously displayed; they are certainly an inch in length. And, for another thing, they are obviously the work of an amateur—scratched by an unpractised hand with a rough and ready tool. A tradesman would have made a better job of it."

" And what are the initials ? "

"There are two—R. F.; there's no mistaking what the letters are, although, as I say, the execution is so clumsy."

Blaise Polhurston appeared to ponder.

"'R. F.'—'R. F.'!—I wonder who's 'R. F.'!"

"That's what the police are wondering. They are inquiring for a man who is known to have had a revolver in his possession, and whose initials are 'R. F.' When found they'll have twisted the first strand in the rope which may ultimately hang him."

The Gentleman leaned back again in his chair. His eyes returned to the fire.

"Baynes, what's your opinion of circumstantial evidence?"

The lawyer stared, as if the query had taken him aback.

"Mr Blaise! How can I tell you in a sentence? I'm not a digest of concretions."

"I'll put it in another way. Supposing this R. F. is found, and certain other apparently corroborative facts are brought out in evidence against him, would you, as a juryman, pronounce him guilty, although no one pretended to have seen him kill the man?"

"If visual evidence of the actual killing was necessary for a legal conviction, then I should say that the law is a greater ass than even popular prejudice supposes it to be. Such a ruling would let loose on society ninety-nine of every hundred notorious assassins."

"Then you would pronounce him guilty?"

"If the revolver is shown to be his, and he is not able to give a satisfactory explanation of how it came to be discovered where it was; and he is found to be, or to have been, in possession of the missing property, and cannot account for how he came by it; then I, for one, should require little else to induce me to consign him to the fate which I should be convinced that he fully deserved."

"That is odd!"

"Odd! How odd? Do you mean to say that you yourself, the circumstances being such as I have described, would not be of opinion that the presumption of guilt was overwhelmingly strong?"

"I suppose I should be, which makes it odder still."

"I don't understand you, Mr Blaise. Your capacity for what I call hair-splitting appears to be in advance of mine."

"You think so? That you should is perhaps the oddest part of it all. Come, Baynes, you must bear with and forgive me.—So no one heard the shot?"

"Not a creature. What makes that more remarkable is that Lucas' room immediately adjoins; that he was in it at the time, and that the discharge of such a weapon makes a considerable noise."

"So Lucas was in his room. Was he looking out of the window?"

"Looking out of the window? What do you mean?"

"As the—murderer passed along the balcony."

"Balcony? What balcony?"

Mr Polhurston cogitated. That odd smile came back to his face.

"You see, Baynes, it's this way. Since I heard of Shapcott's accident—"

"Murder!"

The correction was accepted.

"Murder, I've been constructing, as it were, some little theories of my own. I went one day to view the premises externally. Someone pointed me out Shapcott's room, and I noticed that what seemed to be a narrow balcony ran past his windows along the entire block; so I asked myself, in my search for

H

a theory, supposing the miscreant escaped along the
balcony, and Lucas was standing at his window, with
the blind drawn half-way up, would he, from inside
the lighted room, have seen him pass?"

"I haven't the faintest notion."

"He has hinted at nothing of the kind—or at
having his attention attracted by anything?"

"Certainly not.   I have myself questioned him
more than once; he asserts that he saw and heard
nothing."

"How did the villain gain access to Shapcott's
chambers?"

"That is more than we can learn.   The porter says
that he never left his post in the hall, and yet he
declares that no one to whom suspicion can in the
slightest degree attach passed either in or out."

"No one?"

"No one!   As you have seen the place you will
know that Shapcott's rooms are at some distance
from the ground.   It would have been impossible to
climb up to them, if anyone succeeded in climbing
down.   And, if anyone did climb down it was at the
immediate risk of breaking his neck."

"It is something of a drop."

"Drop!   To drop would be certain death."

"You think so?" Mr Polhurston rose from his
chair.   "Well, Baynes, it seems to me that about
this business there's a flavour of mystery.   Were I
upon a jury appointed to try a man accused of
killing Howard Shapcott, I fancy that I should want
something more than circumstantial evidence before
I would agree to let him hang."

"What more would you want?"

"I should require proof, Baynes, actual proof."

"What sort of proof would you require?"

"Ah, that's the question!   What sort of proof
should I require?   I wonder what proof would con-

vince me that a certain A or B was Shapcott's murderer ? "

The lawyer turned away with an impatient gesture.

" If all men were like you, Mr Blaise, none of our lives would be safe."

CHAPTER XI

THE TAILOR MAKES THE MAN

Mr Polhurston regarded the lawyer with a curious scrutiny, fixing his eye-glass, with its grotesque attachment of coloured twine, in his eye, the better to enable him to do so. He laughed, as in the enjoyment of a jest which had a sardonic flavour of its own.

"Well, Baynes, let's cease to talk of 'grinning murder with jaws all bloody.' To change the theme, I'm a man of fortune! Do I look it?"

The expression of Mr Baynes' countenance, as he surveyed the tatterdemalion figure in front of him, conveyed a negative with sufficient force.

"We will soon change all that, Mr Blaise, and, if you will take my urgent advice, at once. I cannot but think that you incur a serious risk by remaining so long in those damp clothes."

"But, as I have explained to you, when one is used to running risks, risks are nothing. Still I shall have pleasure in acting on your friendly recommendation, but—with twopence in the lining of your hat!"

"You can have any sum which you require."

Mr Polhurston began to swing his eye-glass, pendulum-wise, to and fro, regarding it with apparent absorption as it swung.

"You say there is £2489, 16s. 3d. at the bank?"

"Those are the figures."

"Give me a thousand."

"When?"

"Now."

The lawyer appeared to deliberate, which seemed to afford the other some amusement.

"You hesitate. May I inquire why if your tale's all true?"

Mr Baynes spoke as if experiencing a difficulty in finding the most fitting words with which to clothe his thoughts; with a certain timidity, as if fearful to wound, yet desirous to impress.

"I hope, Mr Blaise, that you won't forget how eagerly you are awaited at home."

The Gentleman continued to swing his eye-glass.

"My dear fellow, don't you trouble yourself about what I remember or forget."

There was a pause.

"I hope you are not going to disappoint them?"

"'Hope on, hope ever!' The thousand, Baynes—the thousand."

"I don't wish to offend you, Mr Blaise, by showing a curiosity which you may esteem impertinent, but won't you give me some inkling of your plans?"

"Plans? My plans are to get food and a change of clothing. Just now you urged me to make haste about the latter, yet now you keep me dallying."

"You don't need a thousand pounds for clothes."

"Baynes, this fortune, of which you have been telling me so much, is it yours or is it mine?"

"Yours, beyond a doubt."

"Then give me that thousand pounds."

He looked at the lawyer with a smile which sent him, without another effort at expostulation, to his desk.

"Shall I give you an open cheque, or the cash?"

"The cash, my friend, the cash! Cheque me no cheques! And let it take the shape of eight one hundred pound notes, ten tens, ten fives, and the rest in gold; with perhaps a modicum of silver."

The lawyer, having written out a cheque, withdrew with it into an adjoining room, and presently returned.

"I trust, Mr Blaise, you will at least do me the honour of taking lunch with me."

"A thousand thanks, but I can't."

"Then you will be my guest at dinner?"

"I would if I could, but it's impossible."

"When shall I see you again?"

"I'll drop you a line—I'll drop you a line."

"Does that mean that I shall not see you again for an indefinite period?" The other shrugged his shoulders. "Answer me, Mr Blaise."

"It means that when you will see me again, or under what circumstances, is more than I can tell you."

"Why not?"

"I'm under no compulsion to assign you reasons, Baynes."

"But there are a hundred points on which I wish to consult you, which require your immediate attention, and on which I cannot move without your sanction."

"They must wait."

"What is your address?"

"I've none."

"I presume that you will write to them at home?"

"Presume."

"Yes or no? If you knew with what an agony of longing your mother looks for you, you would not hesitate."

"I'm no letter-writer, Baynes."

"Then what am I to say when I write?"

"What you please."

"I do entreat you to let me know when I shall see you, or hear from you again. Do not make use of this money to drop back again into space. Remember how deeply I am engaged to your mother to leave no stone unturned so that she may fold you in her arms again before she dies. To be plain with you, Mr Blaise, your manner fills me with all sorts of fears."

"Fears are curious cattle. One knows not whence they come, nor why, nor how. Have you observed it, Baynes?"

They continued chopping phrases, the lawyer pressing questions on the Gentleman, endeavouring his utmost to extract from him some definite undertaking; the Gentleman fencing, parrying the most pointed thrusts with a zest which suggested positive enjoyment of the man-of-law's anxiety. They were still at it when the clerk entered with the change. Mr Baynes handed it over to its proper owner, counting it as he did so, with an air of reluctance which more than hinted that he would have detained it if he could. Mr Polhurston, on the other hand, received it with unconcealed delight, handling the bank-notes with an appearance of childlike satisfaction, which went oddly with the manifold marks which he bore on his face and his attire of a prolonged wrestling with the world, in which he had endured many a fall.

"A bank-note for a hundred pounds! Ye whales and little fishes! Do you know, Baynes, that a generation has grown up since I've so much as looked on one—and here are eight of them—and all of them my own. And all that heap besides—that pile of yellow-boys—pretty, rounded, shining, little things. How they gleam — and how they tinkle. To feel what I am feeling, it's almost worth while to have gone penniless for a hundred years—to have again a handful of money, and such a handful! It makes my blood glow in my veins—and just now, I had no blood to glow."

The lawyer was sententious, and trite.

" Money is a good friend, but a bad enemy."

" Enemy!   How can it be an enemy ? "

" Easily ; when its allurements cause us to neglect the natural claims of kinship."

This time the lawyer's tone was dry. His companion laughed.

" Ah, Baynes, you harp upon one string with a persistence which I find inartistic. Till our next meeting ! "   And the Gentleman was gone — with the thousand pounds.   Mr Baynes, after his departure, gave audible utterance to a few sentences, which were, possibly, intended to relieve his feelings.

" As opinionated as ever and as quick to resent advice. Suffering seems to have taught him nothing ; not even a son's duty to his mother. What shall I tell that anxious-eyed woman down in Cornwall, who always looks out of the window, and whose ears are continually strained to catch the sound of footsteps, which once she knew so well? That I found her son in a ditch, that he climbed painfully out of it, to snatch at the money which I offered, and then climbed back again without delay? A pleasing tale to carry to the mother, to whom pride has already brought destruction."

The suspicion of sleet had become a certainty. A mixture of snow, hail and rain was being driven before a blustering wind. The Fields was a wilderness. The Gentleman, who had refused Mr Baynes' offered escort to the door of his chambers, stood looking out upon the scene, with sudden recollection that he was hungry and that his clothes were tatters. As he stepped out on to the pavement the gale caught his hat ; he only just snatched at it in time to prevent its being blown into the road. A hansom passed. He hailed it. The driver stopped.

" What's up ? " he asked.

"I want you."

"What for?"

"What for, man!   I suppose you've plying for hire. I'm a fare."

The cabman was derisive.

"Oh, you're a fare, are you?   A pretty sort of fare you are.   One of the Royal Family, ain't you?   Do you want Buckingham Palace, or is it Marlborough House you're stopping at just now?   If I was to take to driving blokes about like you I should make a do of it, shouldn't I?   It's Black Maria you want, cocky.   So long!"

The cab drove off, the driver saluting him with his whip as he went.   The Gentleman was conscious of making an effort to appreciate the fellow's humour with but imperfect success.   He put his hand up to the inner pocket of his coat, in which his treasure was deposited.   It's touch reassured him.   A half-fantastic fear had flitted across his mind, lest its presence had been but part and parcel of a dream, and that, after all, he was, in reality, still the mendicant he seemed. The crinkling bank-notes, the solid coin, restored his courage.

The next time a cab approached he tried another method.   He held up half a crown.

"Drive me to Ludgate Hill for that?"

Pulling up, the cabman regarded him and it askance.

"What is it?   Let's have a look—half a dollar? Is it a good one?"

"To the best of my knowledge and belief."

"Where did you get it?"

"That's my affair."

"Pay in advance?"

"With pleasure,   I am willing to trust you."

"Jump in!   Mind you sit as far back as you can; if anyone sees you in my cab they won't care to get in when you get out."

The Gentleman got in; the driver let the shutter down, then addressed his fare through the trap in the roof. "What do you want at Ludgate Hill?"

"To enjoy the air."

"Gay sort of air to enjoy, this is, ain't it? What's the game? Coppers after you for pinching that half-crown, along with one or two more on top of it? Doing a bolt, are you?"

"I wish you'd do a bolt, or your horse."

The cabman laughed; he shut the trap, and took the hint.

At the foot of Ludgate Hill the fare alighted, the hansom dashing off as if ashamed of having had even momentary connection with such a figure of fun. The Gentleman entered one of those clothing establishments in which a man can obtain all the varieties of apparel which he needs. A shopman advanced.

"Well, my man, what is it?"

"It's a customer."

"A customer! You? What do you want?"

"Clothes. Don't you think I look as if I did?"

The shopman thought so so strongly that, there and then opening the door, he waved his hand towards the street with a gesture which was more than suggestive.

"Now, my man, out you go; we've got nothing in your line to-day. You'd better try elsewhere."

The Gentleman resolved to ride the high horse. This person required crushing. To be refused as a fare by a cabman was bad enough, but for this individual to decline to supply him with the articles which would immediately transform him into another being, one who would have honour in the eyes of all men, and who would be free from the repetition of such insults, that was not to be endured.

"Send me your master, sir. You are evidently incompetent to fill the position which you are supposed to hold. Is that the way in which he has instructed you to attract customers to his establishment ? "

The man hesitated, evidently struck by the difference between the speaker's tone and his appearance. Someone else approached—a shopwalker.

"What does this person want ? "

"He says he wants some clothes, sir."

The shopwalker turned to the Gentleman.

"What clothes do you wish to buy ? "

"An entire wardrobe. I have sufficient money, though I admit that you are justified in doubting it. My name is Blaise Polhurston, and I have just received a payment from my solicitor, Mr Henry Baynes, of Lincoln's Inn Fields, to whom, if you like, you can refer while I am selecting the goods which I require."

He held out a packet of bank-notes and a handful of gold. The sight was sufficient. He was shown into a private room—his appearance was a trifle too conspicuous for the general shop—where he received all the attention he could possibly desire. His purchases were lavish; he provided himself with an entire outfit. When he had transferred himself into a portion of it, the assistant, who had been waiting on him, stared. The transformation was almost ludicrously complete; he had become so unmistakably what the world calls a gentleman.

"Excuse me, sir, but I shouldn't have known you."

"Clothes do make a difference; read Carlyle. But a taste of experience is more education than even the Sage of Ecclefechan's theories. Have you tried what you looked like in a costume of ill-assorted rags and tatters ? "

The assistant was a well set-up young fellow, not bad-looking; he involuntarily put his hand up to his collar to adjust his necktie.

"Not yet I haven't, sir; and I hope I never shall."

The Gentleman laughed.

"Well, as I say, it's educational; but, perhaps, your education's finished."

"I don't know about that; but I don't mind confessing that I do like to see a man dressed well—especially if he is a gentleman."

The Gentleman laughed again. In the new position of affairs he found the fellow amusing.

To the rest of his purchases he added the necessary trunks to hold them. When he quitted the establishment he had an obsequious and grateful shop-walker as an escort. A hansom awaited him without, whose driver evinced no symptoms of a disinclination to accept him as a fare. He drove to a hotel, where, having engaged a comfortable room, and established himself therein, he treated himself to a bath. The bath added another touch to the metamorphosis. He emerged from it another creature. The man who was had gone; if the vein of luck, on which he had stumbled, endured, that man was gone for ever. But lunch was needed, and a few more oddments for the miracle to be rounded off as became a workman.

He went from the hotel in a cab to be shaved; another touch. Proceeding thence in another cab—a well-dressed man, although provided with a capacious umbrella and luxuriously overcoated, could, of course, not be expected to go a step on foot on such a day—to one of the so-called "stores" where everything is sold. There he bought a watch and chain—an excellent watch and an elegant chain; a gold sovereign purse—it is the only kind of purse a man can carry; a box of really good cigars—Bock, Obsequios del Rey, 1892, 120s.; other essentials for a smoker, and various addenda, which he had sent on to his hotel. Having completed which arrangements, he judged himself to be fairly entitled to consider the question of lunch.

Which was not a matter to be treated lightly. He was conscious that, as a man of fortune, he might never again, in the whole remainder of his life, be in possession of such another appetite as he had then. That the capacity for enjoying good and well-cooked food would probably continue for, at least, some time, was possible. After going so long hungry it was hardly likely that he would at once reach the seats of the sated. But the meal of which he was about to partake would be of the nature of a function; it would mark the date of his hegira —of his being born again. The thing certainly amounted to re-incarnation.

Running over a list of restaurants in his mind, he selected one—one of the most famous in town. How long ago was it since he stood outside its doors, pretending to sell matches, looking out for a chance of a copper? He had not dreamed then that he would ever cross its threshold as a guest. Was it not last week? How odd! It seemed a century ago—already. He directed the driver of another cab to take him to that palace of high feeding. The lunch should be a royal one.

As he went he drew up the menu—he flattered himself that he had not yet forgotten how to. The gourmet which was in him was, even yet, not quite atrophied; he would prove it. The meal should be unsoiled by indulging any tendency towards looking back or looking forward. A mind at rest was necessary for its appreciation—he would have it. All his forces should be concentrated on the enjoyment of the moment, and on that alone. It should be a feast of the gods.

As he got out of the cab, a man came up to him who was selling newspapers — an elderly man, in whom he seemed to see a colourable imitation of what he himself had been so short a time ago. In front of him the man held a placard. The words on it caught the Gentleman's eye:—

EMBANKMENT CHAMBERS MURDER.

———

Accused at Bow Street.

CHAPTER XII

THE GENTLEMAN AS A GENTLEMAN

FOR a second or two Mr Blaise Polhurston stared at the placard, as if with a total lack of comprehension of what it conveyed. The cab was at the back; the restaurant porter at his side; the newsvendor accosted him in front. But during those few seconds he saw neither of the three. His eyes were with his thoughts, and they were at some distance from his body. The newsvendor's voice recalled them with a sense of shock.

"Buy a paper, sir? Fifth edition—*Evening News, Globe,* or *Star?*"

The man held out all three. Mr Polhurston took them, handing him a shilling in return.

"Never mind the change."

"Thank you, sir! Gawd bless you, sir!"

The fellow's voice quavered. Mr Polhurston doubted if he would have used precisely that phrase, though he thought it probable that his gratitude might have been quite as keen had such a windfall come his way —a hundred years ago.

He advanced towards the building. The porter spoke to him with uplifted finger to his cap. "Cabman wants to know if he's to wait, sir?"

He remembered then that he had not paid his fare.

"No. Give him that."

He entered the restaurant. It was well filled, pos-

sibly by people who felt that on such a day to eat
and drink was best. He selected a table. A waiter
came to him.

"Give me the *carte*. I'll call you when I'm
ready."

The man, laying the bill of fare in front of him,
withdrew. Mr Polhurston sat quite still, looking
neither to the right nor left. He knew that he had
had a narrow escape. The foundations of the world
had trembled beneath his feet; a little more and they
would have slipped from under him. Even yet he
had to grip tight to keep his hold on life. It seemed
all at once to have become fluid.

He held the three papers in his left hand. A
crinkling sound recalled them to his recollection. It
was they which had been responsible for the mischief.
That contents bill had brought him back to earth
with a bump. How did it run? "Embankment
Chambers Murder. Accused at Bow Street!" The
words danced before his eyes in the vacant air, as if
he could not dodge them. "Accused at Bow Street."
What did it mean?

He brought the papers down upon the table with
a sounding thwack, startling the other lunchers. To
ask himself such questions at such a moment would
be to ruin that luncheon of the gods.

At least let him derive the fullest possible amount
of enjoyment from what he was about to eat and
drink, if to-morrow he had to die. Who was the
"accused" who had figured at Bow Street? Who
could it have been? What an ass he was to ply himself with such questions at that supreme moment of
his re-incarnation.

"Waiter, bring me a liqueur brandy!"

He was sufficient of an artist to be aware that
cognac was hardly an appropriate prelude to a perfect luncheon, either as an appetiser or as a cleanser

of the palate. It was the fire in the stuff which he
wanted, the force. When the waiter brought it, he
swallowed the tiny glassful with a sense of grateful
refreshment. It seemed to reach his nerves, to brace
them.

It was a royal meal which he commanded—requir-
ing, at that hour of the day, a hearty appetite for its
right appreciation, but, given that, as good an one as
one might wish. He began with an oyster or two;
the soup was a bisque; the fish was a grill; the
entrée slight; the roast a woodcock.

With each course he ordered a separate wine. There
was a slight movement of the waiter's eye at this.
He glanced at the customer out of the corners. A
single wine has become so general—except at livery
dinners, and similar feasts of Gargantua—that to find
a man ordering half a dozen even small bottles of
different sorts for his own consumption is surprising,
though he may only propose to drink a glass out of
each. As a matter of fact, Mr Polhurston intended
to do more. He meant to drink if he did nothing
else, for, oddly enough, he was doubtful if, all at once,
at the psychological moment, his appetite had not deserted him.

Perdition seize those papers! Why had his glance
been so unfortunate as to light upon that miserable
placard? What a difference the chance encounter
had made in the world's appearance. He thrust the
printed sheets resolutely from him to the extremest
limits of the table. He would not see what was in
them till he had lunched—he would not. His fingers
itched to open them; his eyes yearned to run up and
down the columns in search of he did not care to
think what; he was conscious of a morbid mental
longing to learn what there was to know—to learn
it all. But he put a strong restraint upon his im-
pulses, vowing that he would not so much as peep

I

within their pages until he had discussed the menu and, at anyrate, tasted of every dish.

Appetite comes with eating. So soon as the food began to cross his lips there came with it the capacity of enjoyment as a sauce. He ate a hearty meal, doing full justice to every dish, after all. Good food tempts an unfed man even when he's standing, and knows that he is standing, at the gallows' foot.

"It's something to eat I wanted," he told himself, as he became conscious of the process of building up which was taking place within. "The English cannot fight on empty stomachs—the story's old. "'Feed us and we're a match for a whole world in arms.'" He drained his glass; he was doing full justice, also, to the wine. "A man's mental equipment cannot stand alone, it needs a physical basis. Unstoked furnaces won't drive a ship through the calmest sea, but let them be well coaled they'll take it through the wildest storm."

He laughed to himself with positive gusto. Leaning back in his chair he glanced about him with a very real consciousness of life's good things. Was it the Fat Knight who had said that if his belly was well lined, and his throat well oiled, nothing else much mattered; or was it someone else? In any case the man's philosophy was sound beyond a doubt. What did it matter if to-day one could eat, drink, and be merry whether to-morrow one died?

A voice addressed him from behind.

"Polhurston! Surely it is you!"

A hand was laid upon his shoulder. Startled, turning, he saw a face looking down at him which was like a breath of air from the days which were past. It was a handsome, soldierly, strong face. The flowing moustaches were just turning grey. In the bright eyes there was a glint of boyish laughter.

"Max Liddell!'

The new-comer grasped Mr Polhurston's hand in both of his.

"It is you. I knew that it was you, I could have sworn it. My dear chap, this is a stroke of luck. To think that you should have been here to-day! I wouldn't have missed seeing you for anything. Where have you been hiding yourself all these years?"

The stranger drew his chair up to Mr Polhurston's table. He was not only well-dressed, but he conveyed the impression of never having been anything else his whole life long. He radiated vitality; his looks, voice, manner, everything about him suggested buoyancy. His delight at the meeting was unfeigned—it was boyish.

The Gentleman eyed him inquiringly. He observed appreciatively certain details in his dress, noting how and where he himself fell short. After the society with which he had mingled, one needed to be able to study a model before, in the matter of costume, one could achieve perfection. It was a second or two before he answered.

"I have been wandering to and fro upon the face of the earth."

"No—have you? Jolly, that sort of thing. I've been vegetating these seven or eight years—makes a man feel rusty. I envy you."

If he felt rusty, he did not look it. And as for envy, the Gentleman smiled.

"Do you?"

"I do—really. I don't mean to say I haven't had a good time—a first-rate time; I'm married, you know. You married yet?"

"No, not yet."

"Ah, there's more to be said for the state than some who fancy themselves wise will allow; and I've got chicks. Still, when a man's done a lot of knock-

ing about, sometimes he does miss the taste of the
salt, especially when he meets another chap on whose
lips it's still fresh. Anyhow, you've come back just
in the nick of time. You know that Glenthorne's
going to resign ? "

"I'm afraid I don't—nor who Glenthorne is ! "

"My dear chap, he's our member—Philip Glenthorne, of Saltash. You must remember Glenthorne ? But he's getting on in years—gout's pinned
him ; he can scarcely move from his chair. So he's
going to retire. And when I heard what had
happened, I said to your mother "—Mr Polhurston
winced at the allusion to his mother, though the
other did not notice it—"that you're the very man
we want—and you are."

"For what ? "

"For the division ! With us, Polhurston of Polhurston is still a name to conjure with—you'll have
a walk over. I won't go so far as to say that you'll
be supported by both sides, but it's my firm conviction
that you'll be unopposed."

Mr Polhurston smiled; the idea of his being returned to Parliament had about it such a flavour of
the grotesque.

"Liddell, you move quickly."

"Not a bit of it; and I can assure you that I know
what I am talking about. By the way, I have to
congratulate you. I hear that Shapcott's left you
everything."

"He has."

Colonel Liddell looked at the Gentleman with a
smile which was meant to be knowing, but which the
Gentleman declined to understand.

"Well, you know what I mean, and I'm sure I'm
delighted. You don't, my dear chap, need me to tell
you that ; but one would have hardly expected Shapcott to make you his heir, now, would they ? " The

Gentleman was still. "Unless he meant to be consistent, and continue surprising people with his little eccentricities to the very end; eh?" The Gentleman said nothing. "That was very sad about his death, wasn't it? Though I don't know that I shouldn't be as willing to go with a bullet through my lungs as in any other way."

"If one has to go."

"Precisely—if one has to go. And we all do have to go, that's the point—a little sooner or a little later, eh? For my part, when I do go, I pray that I go quickly; I'd rather face a gun barrel than a surgeon's knife. When are you going home?"

"I have not decided."

"You haven't been yet?"

"No; not yet."

"Then—you won't mind my speaking plainly—go to-morrow." Leaning over the table, the speaker lowered his voice. "You know what a woman your mother is, what a front she carries to the world, how she'd sooner die than own to a wound. I'm confident —my wife tells me, and I have seen it for myself— that the only thing which has kept her alive, has been the hope of getting you back again. The truth's been made more obvious by her desperate efforts at concealment; you shouldn't have been away so long, you really shouldn't. You won't mind my speaking plainly, will you, Blaise?"

"Not a bit. I have been used to the plainest of plain speaking on all sorts of subjects for some years." There was that in the speaker's tone which sounded a note of warning to the Colonel's not over keen perception. He got up.

"You must dine with us to-night. I'm sorry to say that I can't stop now, but there are a thousand things about which I wish to talk to you. You must come to us to-night in Brook Street; we shall be alone."

"Thank you, but I can't."

"Then I tell you what I'll do. I'll come to you for half-an-hour. Where shall I find you?"

"At present I haven't even an address."

The Colonel's face was clouded.

"I hope you're not going to keep off your old friends."

Mr Polhurston seemed to be considering. He drummed softly with his finger tips upon the table.

"As I've said, you move quickly. It's only within the last twelve hours that I have heard that I am Shapcott's heir. You must give me time to accustom myself to the novelty of the situation."

The Colonel was looking towards the door. A lady had entered, apparently acccompanied by half-a-dozen gentlemen.

"Hollo! There's my wife! And Ferrars and Havisham, and a whole crowd." He signalled with his hand. The lady advanced—her escort in a sort of *queue* behind. "Agatha, I think you know Blaise Polhurston?"

The lady stopped, and stared.

"Blaise Polhurston! You! Through what hole in the skies have you dropped?"

"I've popped up, Mrs Liddell."

"I hope you don't find the change in temperature too refrigerative. I note the 'Mrs Liddell!' Well, so long as you don't expect me to call you Mr Polhurston. Have you forgotten Bruce Ferrars and Franklin Havisham?—and—everybody else?"

At once Mr Polhurston found himself the centre of a little lively crowd. Hands clasped his warmly; they slapped him on the back, caught him by the arms and shoulders. Congratulatory words were ringing in his ears. All sorts of pleasant things were being said. It seemed impossible to doubt that the delight of these well-dressed, well-bred people, at seeing him was real. And yet, to him, the whole thing seemed to be as the warp and woof of some fantastic kaleidoscopic dream.

Presently Colonel Liddell's voice became audible above the others.

"I've just been asking Blaise to dine with us to-night, but he won't."

The lady addressed the Gentleman.

"Why won't you?"

"I can't."

"Why can't you?"

"I hope to have that honour on some future occasion."

"Honour! And some! Why won't you enjoy that honour to-night?"

"I repeat—I can't."

"Will you breakfast with us in the morning?"

"I'm afraid I cannot promise."

"What can you promise—to lunch with us?"

"All my arrangements are still in a state of nebulosity."

The lady eyed him with intention. She turned to the others.

"Good people, be so good as to go over to that window until I call you." The escort withdrew with her husband.

Mrs Liddell addressed the Gentleman with what perhaps was meant to be regarded as severity.

"I believe, Blaise Polhurston, that you and I were as brother and sister."

"Once—long since."

"It's not very polite of you to remind me of the flight of time, but anyhow the fact remains, and I intend to assume that that supposititious relationship still exists. Why can't you dine with us to-night?"

"I can't."

"That's a phrase which means anything or nothing. I wish to know just what it does mean, please!"

"Simply that for a great many years I have lived in a different world to yours—a very queer world you would think it—from whose shackles I cannot shake myself free all in a moment. You must give me time."

"How much time?"

The Gentleman's face was illumined by his whimsical smile, in which was such a suggestion of pathos.

"Ah! that I cannot tell you."

Mrs Liddell searched him with her glasses.

"Have you acquainted your mother with your re-appearance on the earth's surface?"

"A thousand pardons, but — my mother is a subject which I must entreat you to excuse my discussing."

"What, again, do you mean by that?"

The whimsical quality of his smile became more pronounced.

"May not a man keep his mother to himself?"

This time the lady showed distinct signs of being huffed.

"Oh, certainly. Pray, what address will find you if I should happen to wish to shed on you a card!"

"You know Baynes, of Lincoln's Inn Fields?"

"Oh, yes, I know Baynes, of Lincoln's Inn Fields—rather."

"For the present, address everything which is intended for me to him, if you don't mind."

"Not in the least. Shall be charmed. One always does like to leave cards for a friend at his lawyer's office. Good-day."

The lady sailed away, her head a little in the air, her plumes ruffled. She joined her husband and her friends. She said something to them. Then, in a body, they left the room, the men waving farewells to Blaise Polhurston as they went.

He remained standing, for a moment, after they had gone ; then, sitting, he gazed, as in a brown study, at the tablecloth.

"She always had a temper," he told himself. "Even when, in her moments of exasperation as a small child, she used to try to tug my hair out by the roots."

His brown study continued for a while. Then, unfolding the *Evening News*, some headlines in it caught his eye,—

"THE EMBANKMENT CHAMBERS MURDER.

"An Arrest at Last.

"Prisoner Before the Magistrate.
"Scene in Court."

Before he could read further, without the slightest warning, Mrs Liddell re-appeared at his side. Her manner was scarcely dignified. Her excited words were uttered in a low and rather tremulous voice, which was intended to reach his ear alone ; her remarks were a little incoherent.

"Blaise, I don't care how you speak to me, or think of me, or treat me ; but I'll tell you plainly that if you don't take the next train, and gladden your mother's eyes by the sight of you, you'll do the wickedest thing that ever you did in all your wicked life. God's curse will light on you, and nothing you do will prosper. There, now, I've spoken my mind !"

As if the fact relieved her, or as if fearful of the consequences which might follow, she retired from the room as unceremoniously as she had entered it. Mr Polhurston stared after her for a moment, in seeming bewilderment.

Then his glance returned to the paper.

CHAPTER XIII

THE REPORT IN THE PAPER

THE headlines seemed to have for him a singular fascination; it was as if he could not escape from their ostentatious inkiness. For more than a minute he continued to stare at them, repeating them over and over again to himself, like the imperfectly educated person who spells his words out, one by one, so that his eyes may have the assistance of his ears.

The afternoon, by now, was well advanced. The lunchers had gone. He was the only one who remained. With the exception of the attendants the room was empty. His own particular waiter was hovering round with an air of suggestion which there was no mistaking; it was plain that he did not intend to allow his dilatory customer to remain to read where he was, if he could help it. Mr Polhurston bowed to the inevitable.

"Bring me my bill." It was brought. "Where can I smoke?"

"I will show you, sir, if you will step this way."

Gathering up his journals, Mr Polhurston stepped that way. In a smoking-room easy-chair he recommenced his reading.

He was calmer, his pulse steadier, his mind clearer. The disturbing influence of the unexpected encounter with the Liddells and their friends had almost passed away. He was better able to grapple with the situation as it really was; with whatever phase it had suddenly assumed. He glanced up and down the column. The report was punctuated, wherever possible, with the unescapable headline. One of them struck him in the face—as it seemed—an actual blow.

"Pollie Hills Speaks Out."

His self-control was less than he had supposed. At the sight of the familiar name his heart stood still. He had to gasp for breath. Others conveyed to him an almost equally lurid significance.

"A Ring and Bracelet." "A Try for
the Reward."
"The Dead Man's Property."

He had to put the paper down again upon his knee, his hands had all at once become so tremulous. He leaned his head back, and closed his eyes. It was a nuisance that he should, physically, be such a fool, that his nerves could be so easily set racketing. If he could not meet a newspaper allusion with apparent unconcern, then he might as well go at once and throw himself over the bridge into the river. He ordered another cognac. The wine he had drunk did not seem to have had the least effect. He would go on drinking brandy till it had brought him to the sticking point.

When he did commence, really, to study the report, it gripped him by the throat, so that he had to read it, from the first line to the last in, as it were, a single breath. He read, so to speak, with his whole soul and body; yet, when he reached the end, he was wholly at a loss to understand the meaning of what it was he had been reading.

"What does it mean?" he asked. "What does it mean?"

He started afresh, more slowly this time; endeavouring, as he went on, to assimilate the purport and intention of each separate word.

It was one of those pictorial reports with which a certain sort of evening paper provides its public. Part of it was from imagination, part of it described what the reporter saw from his particular point of view, and part of it was a transcript, more or less verbatim, of what had actually been said in court. Although Mr Polhurston might be supposed to know something of the subject under treatment, yet the journalist's agility merely puzzled him; and it was from what he had not reported that he had to find a key which would give the whole thing meaning.

The report started, as "descriptive" reports are apt to do, with a flourish. It began by saying that the police had after all done something to show that they were still alive. One of these crimes which periodically shock London had remained too long a "mystery." The police had at last done something to justify their existence by bringing, that morning, before the magistrate at Bow Street,

ROBERT FOSTER.

At this point Mr Polhurston stopped to cogitate.

"Robert Foster! Robert Foster! Who's Robert Foster?"

He remembered what Mr Baynes had said about the initials which were on the stock of the revolver which had been found in Howard Shapcott's room, the weapon which, beyond all reasonable doubt, had done the deed; the two damnatory letters, "R. F."

The reminiscences started a train of thought in Mr Polhurston's mind.

"R. F. Robert Foster? Why, the initials fit the man! What—what an extraordinary coincidence! Suppose—but the thing's impossible. I wonder who Robert Foster is? I've never heard the name before, so far as my recollection serves. But, in the society in which I've lately mingled, one knew one's acquaintances chiefly by names which were not theirs."

He read on, in search of information.

According to the reporter, Robert Foster was, in appearance, of a type "very familiar in police court annals," whatever that might mean. He had refused, said the same authority, to give his age, but was "apparently between twenty and thirty." He must have presented a sufficiently unprepossessing figure in the dock. "He wore no cap, his hair was in disorder, the end of a red cotton neck handkerchief straggled over his greasy vest, his shirt was unbuttoned at the neck, his face was unwashed, his hands were covered with grime." He looked as if he had "recently been indulging in a rough-and-tumble; by no means the sort of person an inoffensive citizen would care to tackle. Big and burly, his truculent looks were in complete harmony with the rest of him."

Mr Polhurston could think of no one to whom this description applied, even allowing for the reportorial tendency towards the "pictorial."

"I can't think who Robert Foster is; it's extraordinary, because one feels that, under the circumstances, one ought to know."

The newspaper man went on to remark that Robert Foster's manners were quite in keeping. He "looked surlily" about the court, "glared" at the magistrate, "glowered at the witnesses," and "when the principal feminine witness appeared in the box, flashed at her a look of hatred which eloquently suggested what he would like to do to her if he could only get the chance."

The "principal feminine witness" was Pollie Hills.

When the Gentleman had ascertained that fact, he put down the paper, leaned back in his chair, and endeavoured, by a series of mental processes, to arrive at a solution of what that fact might mean.

"What is she doing in that galley? What can have happened after I got into the box? Why is she giving evidence against this man?"

To none of these questions could he find satisfactory answers, nor did he derive much more satisfaction from the remainder of the report.

The police inspector's statement had been brief and to the point. In accordance with information received, he had effected the prisoner's arrest; that was, in effect, the sum and substance of what he said. Pollie's evidence had evidently been the feature of the sitting; it was she who had caused Robert Foster to be arrested.

"But why?" asked the Gentleman of himself in vain.

The reporter was not much more complimentary to the witness than to the prisoner.

"Miss Hills," he wrote, "then got into the box, where she did not seem altogether comfortable. She looked at the prisoner, and the prisoner looked at her, which looks could not truthfully be described as loving glances. Miss Hills is a tall, strapping young lady, with short black hair."

"Short black hair?" Mr Polhurston stopped to consider. How came that reporter to describe Pollie's flowing raven locks, which descended well below her waist, as "short black hair"?

"Although attired in masculine habiliments, she did

not present such an incongruous spectacle as might
have been the case with another lady."

"Masculine habiliments?" He had some faint
recollection of her saying that she would pretend to
be him. How far had the pretence been carried?
What was the key to the puzzle?

"Although the garments were somewhat miscel-
laneous in character, they could hardly be said to
have become the wearer ill, because, to be perfectly
frank, in appearance the lady would make a very
creditable man. When the magistrate asked what
was the meaning of the witness's appearance in the
box in such attire, Inspector Avery declared that
other, and presumably more feminine, garments had
been offered her, which she had refused to don."

Mr Polhurston endeavoured to digest this piece of
curious information before proceeding further.

Pollie's evidence was described at a length, and in
a manner which showed quite plainly that the reporter had endeavoured, in his own way, to make the
very most of it. Actually she seemed to have said
little enough, and what little she had said had apparently been dragged out of her very much against
her will. But the scribe had decorated her statements with such a plethora of comments that it was
a little difficult to make out off-hand which of the
assertions had come from her, and which were original
with him.

"The lady commenced by saying that her name
was Pollie Hills—not Mary. 'I'm Pollie Hills, I
am; I don't know nothing about no Mary.'"

It will be observed that the reader was apparently
supposed to take it for granted that the latter words
were the lady's own.

"She knew the prisoner—of course she did; the
lady looked towards the dock as if she knew its
inmate, if anything, too well. Did she recognise

these? The examining inspector handed her, first a bracelet, and then a ring. Certainly she knew them —he knew she knew them. What did she know of them? Why, she'd had them tied up in her skirt this ever so long.

"'Where did she get them from?' The witness did not answer. Instead she looked again at the prisoner, and again the prisoner looked at her. The mutual inspection on this occasion lasted so long that the magistrate interposed. 'Don't look at the prisoner, witness, look at me.' The lady looked at the bench. The question was repeated. 'Who gave them to you?' 'He did.' With her hand the lady made a significant movement towards the dock. 'You are sure that the prisoner gave them to you?' 'Certain —'tain't likely I'd make a mistake about a thing like that.' 'Did he say where he'd got them from?' 'Not a word.' 'Did you suspect anything?' 'Suspected nothing.' 'When were your suspicions first aroused?' 'Why, last night. I see a bill about the murder, and how those were some of the things which had been stolen from the murdered man—course then I knew.'

"At this point the magistrate asked to be allowed to examine the articles in question, which he did with considerable care, comparing them with the description on the placard to which the witness had referred. 'You are sure,' he asked, 'that the prisoner did give you those things?' 'Certain,' said Miss Hills. 'When did he give them you?' The witness seemed to turn the question over in her mind, then replied without any show of hesitation: 'The day before Christmas Day; I remember it, because he said I might consider it as a Christmas present.'"

Once more Mr Polhurston paused in his reading. They had been given on the day stated, and some-

K

thing very like the words quoted had accompanied the gift. The speech had evidently lingered in the girl's memory. Mr Polhurston guessed for what reasons. With what diabolical ingenuity had she attributed the one man's generosity to the other! He pressed his lips together, and he wondered. Presently he resumed his reading.

"The inspector returned to the charge. The murder of John Howard Shapcott, he reminded His Honour, was committed on December 22. Witness states that these articles were given to her by the prisoner on the day before Christmas Day—that is, the 24th, two days after the murder. When the court had been afforded an opportunity to properly appreciate the ominous significance of the close juxtaposition of these two dates, Mr Inspector suddenly brought the hearing to a close.

"That, he said, was all the evidence which, at present, he proposed to offer. The prisoner had only been arrested an hour or two ago, so that there had been no time to communicate with the Treasury. He asked for a week's remand. On the next occasion the witness, Hills, would again be put into the box, and he would be prepared with other evidence. The remand was granted."

So the report closed.

And for Blaise Polhurston, a period of meditation began. What, now, was the meaning of it all? One thing was clear—Pollie Hills had committed perjury. What had actuated her, or what was at the back of her mind, as yet he did not understand. So far as he could judge, she had actually gone out of her way to lie, which seemed a pity, to say the least. It only made bad worse; complicated an already sufficiently intricate situation.

Who was Robert Foster? And how came he to be arrested? Those were the two matters on which

Mr Polhurston was still completely in the dark. On these two points, to him of the first importance, the report said nothing. Had Pollie, in her desire, at the last moment, to evade arrest, sacrificed someone else? In that case, who? A stranger? Or an acquaintance?

In either eventuality, the outlook was an unpromising one for her. If the man was a stranger, he would have no difficulty, and no hesitation, in making clear her perjury. He had only to open his mouth and the thing was done. The parties would change places; he would step out of the prisoner's deck, and she into it.

Then what course would he, Blaise Polhurston, be compelled to pursue? If she still persisted in refusing to point her finger in the direction in which he very well knew it easily might be pointed, could he sit still and see her suffer? Perjury in a capital case was a heinous offence, especially such perjury as this of hers. He shuddered as he thought of the tremendous penalties which she was challenging.

Suppose Foster was an acquaintance—even a friend; though so far as the Gentleman knew he himself was the only friend she had. The irony of it! The matter would not be bettered for Pollie. He would still be able to establish his innocence with ease; it was not to be supposed that he would, for a moment, allow any consideration of any sort of kind to prevent his doing so. On this occasion it seemed that he had not been accorded an opportunity; before the case came on again the bubble would be blown. Then where would Pollie be?

No, the girl was mad; that was the simple truth—incapable of looking the situation fairly and squarely in the face, or she would not have been guilty of such unutterable folly. In a fit of temper she had made one false step; in her frantic endeavours to retrieve it, she was making half a dozen more, in fresh directions, all much falser than the first. If she had only held her tongue, and kept her temper, how different it might all have been. Mr Polhurston writhed as he thought of it.

He conjured up her face, or, rather, it rose before him, plain and palpable, without there being any need of conjuration. How it had appealed to some chord in his nature, whose existence even he had not suspected—how it still appealed!

Had he sloughed his skin? Had he the chameleon-like faculty of transformation, or seeming to himself to be other than he was? Was he in his proper place as Polhurston of Polhurston — as the Gentleman? How came it that he would have given—well, much, to have had her, at that moment, at his side, in front of him, anywhere close at hand, so that she might be within touch of him, and he of her. He confessed to himself, with a burst of sudden frankness, indeed, with sudden intuition, that Shapcott's money would be worth little to him if she was not to share it.

It was odd. For she was—what she was. He knew it; he was under no delusions. And yet—

He sat for some time in that smoking-room armchair, thinking, with the tips of his fingers pressed tightly together, and his lips pressed tighter still. At the end he had arrived at a determination. He rose with, on his face, a curious fixity of expression, and went out. He took a cab to Charing Cross Station, alighting at the top of Villiers Street nearest to the station yard. There for some moments he stood and looked about him. Then from among the throng of yelling newspaper boys he singled out one. He touched him on the shoulder. The boy's back was turned to him. Supposing it to have been a customer who had touched him, he swung round with a couple of papers held out in his hand.

"*Star* or *Sun*, sir?"

"Ike," said Mr Polhurston, "I want to speak to you. Come here."

The boy stared up, recognising his interlocutor with an amazement which was grotesque.

"Lor' bless me!' he gasped. "If it ain't the Gentleman."

Mr Polhurston drew him round the corner, a little under the shadow of the station wall.

"I want you to find Pollie Hills, and to give her a message. I'll give you four half-crowns if you will. Do you think you can?"

"What? Four half-crowns? Find her? Do I think I can? Blimey, if she's alive I will, and that inside of half an hour."

"Tell her that I want her to meet me this evening, at eight o'clock, at the old place. You understand?"

"Do I understand? What do you think? You want her to meet you at eight o'clock—sharp, I'll say, 'cause gals ain't over and above punctual—at the old place; that's right enough."

"Here's the money; and off you go. And don't forget the message."

"Not much I won't. I ain't of a forgetting sort, and that you know."

The boy sped down Villiers Street, with his papers tucked beneath his arm. He ran up the stairs leading to Hungerford Bridge, Mr Polhurston watching him till he turned and passed from sight.

CHAPTER XIV

THE GENTLEMAN AND THE LADY

" POLLIE ! "

" Gentleman ! "

He had been first at the appointed meeting-place, and had already looked more than once at his brand new watch, wondering why she was delayed. He was beginning to feel anxious. Suppose the lad had not been able to find her; the message had not been delivered; suppose she could not, would not come; suppose—a dozen things. Then he saw her coming towards him, across the badly-lighted road. He knew her in an instant. But she did not seem to know him, for she was hurrying past when he called her by her name.

The name by which she knew him came from between her lips like a cry of surprise. She half-started back, looked up at him in amazement, as if she could not believe that it was he. She repeated his name with, in her voice, what seemed almost like a touch of awe.

" Gentleman ! "

" You are late."

" I—I couldn't get here before."

He noticed the catch in her voice; he himself was conscious of an internal fluttering; as it were, a tightening of the muscles of the heart. The "masculine habiliments" which, according to the newspaper, she

150

had worn in the police court, had disappeared. Instead, she was again attired in apologies for a woman's
garments. She wore no hat, an old checked shawl
was drawn over her head. He carried a cloak over
his arm. He held it out to her, open.

"Put that on."

It was a long theatre wrap, of black silk, lined with
scarlet satin, and edged with fur. She glanced at it
askance—in wonder. It belonged to a class of garment
of which she had only dreamed.

"Gentleman!" she said again, with that catching in
her throat.

Seeing her hesitation, he himself drew it over her
shoulders. It covered her completely.

"Pull the hood up over your head."

Perceiving that again she was loth, or at a loss to
understand what it was he desired, again he did it
for her. The transformation was complete. Where,
before, was a beggar, stood a woman clad in rustling
silks. Only her face was visible through its frame
of fur. As he drew the wrap still closer, Mr Polhurston smiled.

"The disguise is perfect. Your own mother wouldn't
know you, unless she peered close into your face.
Come!"

She shrank back.

"Where are you going to take me?'

"That is my affair; your business is to obey.
Come! quick!"

She seemed to observe, as with a fresh access of
timidity, the new note of authority which was in his
voice—the one change more. He gave her no opportunity to argue, even were she disposed. He hailed a
passing cab. In it they sat, silent, side by side, as if
the novelty of the position abashed them both. Presently he put out his arm, and passing it round her,
drew her to him, and kissed her on the lips. She was

perfectly quiescent, but he felt her tremble, and there was a choking sound in her throat.

"Gentleman!" she gasped.

"I've got you!" he whispered.

He, too, seemed to be troubled by an affection of the vocal chords which impeded his utterance. The cab pulled up at the hotel. She drew back affrightedly.

"What place is this?"

"It's a hotel. I'll tell you all about it when we get inside."

She evinced the greatest unwillingness to enter, showing the wild creature's shrinking terror of the unknown. The sight of people lounging in the hall, well-dressed people, women in shimmering splendour, men in evening dress, added to her fear. He had to hold her arm tightly to prevent her beating an undignified retreat.

"What's the matter with you, child? Surely you're not afraid of being stared at. I thought you were afraid of nothing."

When they reached his sitting-room she was trembling like a leaf. He pretended to laugh at her agitation.

"Pollie, I do believe that after all you are a fool, and you've always claimed to be so wise."

She was crouching against the wall, glaring about her like some wild thing.

"What's this? What have you brought me here for? This is not the sort of place for the likes of me."

"Isn't it? That's a pity. I'm sorry you don't like it, for here, for the present, you're going to stay."

"Stay! What! Like this?"

Throwing the wrap open, she disclosed the rags beneath.

"No, not like that—quite. Just come and see what I have here."

He crossed to where a door led into an adjoining room. She looked after him, still sticking close to the wall.

"Why don't you come? What a deal of telling you seem to need. Must I fetch you?" She went to him, glancing over her shoulder as she moved, as if ready, at the slightest alarm, to take to flight. "What are you afraid of? What a coward you are! I sha'n't eat you, and there's no one else! Surely you're not afraid of me?"

He led her through the door. There, on the bed, and chairs, and everywhere, were a woman's garments—an entire feminine wardrobe.

"What do you think of that? There are enough things to go on with, aren't there?—even to boots and shoes, of which you'll find all shapes and sizes. Those which fit you we'll keep, and those which don't fit you we'll send back again."

She glanced round the room, her eyes seeming to dilate as they passed from object to object. Then she turned her face towards him.

"Who—are these—things for?"

"For you."

"For me!"

The immensity of the statement apparently took her breath away; she was still. He watched, with curious interest, the astonishment which was pictured on her face.

"Don't you want them?"

"Want them! Want them! Where did you get 'em?"

"You forget that I'm a man of fortune. I bought them with the first fruits of my wealth."

"Bought 'em! My! And yesterday you hadn't only twopence!"

"Ah, but that was yesterday; this is to-day. Come out of those things, and into the others; through that door you'll find a bath, only don't be unmercifully long, because when you are ready, you'll find me waiting."

He shut the door; presently she opened it again.

"I say, I've been looking at some of these things, I'm not so sure I know how to put 'em on."

He laughed.

"Ring the bell, and summon the maid; she'll give you all the instruction you require."

Her face darkened.

"Not me! I'm going to have no maid playing tricks with me, and so I'll let you know."

She disappeared, shutting the door again, with a bang.

When, as he thought, he had waited sufficiently long, he tapped at the panel.

"Aren't you ready? I am."

"I sha'n't be more than half a second."

The half-second was, perhaps, somewhat extended, but, in a few minutes, she did appear—in a fashion of her own. Her bedroom door was opened about six inches, and, through the interstice, her voice was heard.

"If you laugh at me, I'll kill myself. Promise me you won't!"

He did laugh, there and then.

"You goose! Why should I laugh at you?"

"You're doing it now. I know I'm a sight, and if you make game of me I'll tear everything right off, straight."

"My dear child, I'll keep the straightest countenance you ever yet encountered; and as for game, I'll none of it."

She hesitated, feeling, probably, that the promise wore an equivocal hue. Then, as if taking her courage in both her hands, she strode hastily into the room; then suddenly stopped, looking at him with defiant eyes.

"Well, why don't you laugh?—split your sides? Go on."

"My dear child, why should I laugh? You look charming. I thought you might be hungry, so, while you've been decorating yourself with your fine feathers, I've made them bring us something to eat, if you're ready."

His manner baffled her. She seemed to search his countenance for the criticism which she defied yet feared. Then her glance turned towards the table, with its white napery, glittering with glass and plate. Her lips twitched at the sight of the food.

"Hungry? I'm starving! I haven't eaten anything since—I don't know when. I couldn't eat what they gave me at the station."

"No, so I should imagine. In such quarters one's appetite would be apt to fail one. Come, fall to." She sat down on the chair towards which he motioned her. "I fancy that, as regards hunger, this morning I was in even worse plight than you are now. I was hungry!"

"This morning? Oh, you—you did get away all right?"

"I look as if I did, don't I? But we'll talk about all that later. At present we'll just eat. It is so long ago since you and I knew what eating really means, that we must stand excused if we make the most of such opportunities as offer."

Although they ate heartily the girl did not display so much voracity as, under the circumstances, might very reasonably have been expected. He noticed, too, that she watched his movements, being careful to do, so far as possible, exactly what he did; and while, of course, there were lapses—she, for instance, handled her napkin very gingerly, as if desirous to neither crease nor soil it, and made a freer use of her knife than strict etiquette demanded—still, on the whole, there was much less *gaucherie* about her bearing than he had expected. At the close she laid down her knife and fork with something of a clatter.

" Well, now, I am feeling more like. That's the best feed I ever had in all my days, I give you my word."

" Yes, and it's a good many days since I've had one as good. It is a comfort to have enough to eat, and to be able to eat it, eh ? And it's one of those comforts which we've had frequently to do without."

She looked at him in silence, her big black eyes seeming to drink in all the details of his attire.

" You're—you're a gentleman."

" So you've always called me."

" Ah, but you're a real one now—a toff, with plenty of money."

" Isn't it possible to be a gentleman, a real one, without the—plenty ? "

" Yes, but there's differences; it's not the same—you know it ain't."

" It's not quite the same, and—I do know it, or, at least, I ought to."

He sipped at his wine, then flirted the glass before his eye.

" You look all right, a-drinking of your wine, all togged out proper; but I never sha'n't—never!"

" What do you mean ? "

" What I say. Look at me—what a sight I am! And I feel it, too."

" I don't know what you feel, but I am looking at you; and shall I tell you what I see ? "

" Oh, go on ! Make game of me."

" Well, to begin with, I'll tell you one thing I see— I see the woman whom I love."

"Gentleman!" The girl's neck and face turned crimson. "Don't—don't you do it!"

"It's true; you know it. You and I have always been tolerably candid with each other; let's continue on those lines. During the course of to-day I've been turning things over in my head, and I've come to the conclusion that I can't get on without you. That's the truth."

"Gentleman!"

She leaned her face on her hands. Getting up, and going behind her chair, he touched her on the shoulder.

"What's the matter with you? I didn't know you were one of the crying sort."

"I'm not! I'm not!" She looked up at him, with streaming eyes. "My Gawd, I wish you'd never had no money!"

"That's kind. I'm afraid that, as at present advised, I don't agree with you."

"It's made such a difference."

"In what way? The chief difference is, that last night we were starving and in rags, and to-night we're well fed and tolerably clothed."

"That's not all. Do you think I don't know? Do you think I can't see that I'm not fit to be in the same room as you now, except as a servant, or something of that? Do you think that I can't see that you're all right in a place like this—just where you ought to be? Am I? Am I? What do you think? Look at your clothes—they do suit you proper. You look as if you'd never worn no other kind. Do I look as if I'd never worn no other kind? Don't I look a proper guy—like a servant girl in her missis's things, or some silly fool who doesn't know no better? If anyone was to come into this room they'd wonder what the dickens I was doing here. They would think no good of me."

"Let them think. It's rather late in the day for you and me to consider what other people think of us."

"Do I look a lady? Do I look a lady? Tell me that."

"You look quite sufficient of a lady for me."

"That's no answer. You answer me. Do I look a lady? You've never told me no lies, so that I know what you'll say; but I want to hear you say it."

"Then you'll have to want. We'll talk about these more important matters a little later on. At present there's a question which I put to you. What's become of your hair?"

She put her hand up to her head with a startled gesture, as if suddenly reminded of its absence.

"Cut it off."

"Did you? Then you didn't do it in very artistic fashion; it's all uneven lengths and jagged ends."

"So would yours have been if you'd cut it off with a knife in the dark."

There was an interval of silence, during which he kept his eyes fixed straight in front, as if endeavouring to resolve a problem which stared at him through open space.

"I see; and I think I understand. And now don't you understand why I am strongly of opinion that you are sufficient of a lady for me?"

She looked at him askance.

"What are you driving at?"

He glanced at her with his whimsical smile.

"I'm driving at something; perhaps you'll understand more clearly what as we go on. Now I want you to tell me what you did after I got into, or you put me into, that box?"

"You tell me what you did."

"I got out again, as is sufficiently obvious, since I

am here. But what you did is not by any means so clear, if I may judge from what I have seen in the evening papers." He held out a paper. "What's the meaning of this?"

He was pointing to a displayed headline—

"THE MURDER IN EMBANKMENT CHAMBERS."

She glanced at it, then turned away.

"Oh! that's what you brought me here for. Now I see."

"Yes, that is partly what I brought you here for. I want to know what it means."

"Find out!"

"I intend to—from you."

"Take you all your time. I'm off. Good-bye."

"Pollie!" He stopped in front of her; she had made a movement towards the door. "Do you really wish to hang me?"

"Hang you!"

"Yes; you will have to be careful if you don't. So far as I am able to judge you are making strong running in that direction."

"Gentleman!"

"I don't know what I have done to you that you should wish to dispatch me to the gallows with so much urgency, but, if you don't, you will tell me at once exactly what this means."

He tapped the paper with his finger. She sat down on the chair from which she had just got up, and stared at him with troubled countenance.

CHAPTER XV

A SINGULAR PROPOSAL OF MARRIAGE

HE asked her the next question while she still stared at him moodily from the chair.

"Who is Robert Foster?"

She answered as before.

"Find out!"

"Is that not precisely what I am doing? Tell me —who is Robert Foster?"

"You know who he is."

"Possibly; but I can't recall him by that name."

"It's Bob."

"Bob!" Mr Polhurston started. "What! The young hawker?"

"Coster Bob—that's him."

"But—I thought he was a friend of yours."

"He might want to be—like his impudence—but he ain't. I give you my word that I'm no friend of his."

"So it seems." Mr Polhurston's tone was dry. "Still, what induced you to make against him such a charge as this? Mere unfriendliness?"

Again he tapped the paper with his finger. Her manner was sullen.

"He shouldn't have interfered."

"How?"

"I dressed up as a man, like I said I would, and

then, when the coppers pinched me, he came up and gave the show away—said I wasn't the cove they was after—said I was a girl. So when he rounded on me I rounded on him."

"In what way?"

"I said that he was the cove they were after, since he was so keen, and it was to save him that I'd togged myself up as a bloke. So they put the buckles on him there and then."

"Pollie!" He turned slightly away. "But he can disprove your statement whenever he chooses; then what will become of you? You can't make false charges with impunity."

"Never you mind what becomes of me; I can look after myself, don't you be afraid. Besides, disproving is not so easy as perhaps you think. From what I hear the rope's about as good as round his neck already."

"Pollie!"

"That's what I'm told."

"It's impossible."

"Is it? Perhaps you'll see. It's not so hard for the wrong bloke to get himself hung as some suppose."

"Then do you seriously suggest that this lad's to hang for me?"

"Why not?"

"And I'm to suffer it?"

"I don't know what you mean by suffering. So long as it's not you, I suppose that's all right. I don't see what suffering there is in getting off."

"That is your point of view—and a not unstriking one. And you're to act the part of the god out of the machine?"

"I don't know what you mean; you talk too fine for me. I got you into the mess, and I told you I'd get you out of it, and so I will. That's all I say and that's all I mean, no more and no less."

L

"And this is your notion of getting me out of the mess? It is not without a spice of originality. And, in the meantime, what am I to do?"

"Do? Why, nothing. You've got lots of money; spend it. You're a toff; be one. I don't know what else you want to do. That's enough, isn't it?"

"You think so? What a—what a high opinion you seem to have of me!"

"What d'you mean?"

"That is the question which I ask myself. What do I mean?" Turning away from her, he looked down into the fire. "Pollie, you seem to take it very much for granted that—I murdered this man."

"Of course; I know you done it."

"Do you? That's frank. Suppose, by any chance, that it should turn out I didn't?"

"No fear! Where did you get those things from? And why didn't you say nothing about it to me? I'm not a silly fool!"

"I see. It's odd how, from given premises, one may draw certain apparently logical, yet wholly erroneous, conclusions. Your point of view, I take it, is the public's. So is justice done. And you are quite clear, in your own mind, that I murdered the man?"

"Certain. Stands to reason, don't it?"

"I suppose it does stand to reason. And you don't think any the worse of me on that account?"

"Why should I?"

"Precisely, why should you?"

"It's no business of mine, only I wish you'd told me at first, then there wouldn't have been all this trouble."

"You think not! I wonder." There was a pause. "Now, what do you propose to do?"

"What do you mean? I don't propose to do nothing."

"Then I do." Turning towards her, he looked at her with his whimsical smile, fixing his eye-glass in its place. "I propose that the case against Mr Robert Foster, *alias* Bob — he always struck me as a very decent youngster, Bob—shall fall to the ground."

"How do you make that out?"

"I trust to be able to make it out by the withdrawal of the principal witness."

"Who's that?"

"You. I propose to take you away with me."

"To take me away with you. Where to?"

She gasped; as if half-uncertain and half-fearful of what it was that he might mean.

"'Over the hills and far away'—ay, and farther still than that. To the end of the rainbow, where the crock of gold lies hidden." He moved towards her. "Polly, I propose to take you away with me as my wife—that is, if you'll be my wife."

"Your wife? Me? Your wife?"

As she had done before, she put her hands before her face. He went closer.

"I had hoped that the prospect of becoming my wife would not have made you quite so sorrowful."

"Your wife! I look as if I was the kind of wife for such as you—and I sound like it, don't I? My God!"

She glanced up at him, with her big, black, velvety eyes, in which was such a conflict of emotions; then covered her face with her hands.

"See, Pollie, it's this way. I don't wish you to appear in evidence against this lad, and I'm sure it is not your own wish. I begin to have a glimpse of how circumstances have been, in a measure, responsible for tangling the matter up, and am convinced that it is not your desire to do the youngster harm."

"I don't say that it is. He's never done me harm, that I know of, only he shouldn't have interfered."

"Well, we'll leave that point alone for the present; because it seems to me clear that, under the circumstances, you will find it difficult to avoid giving evidence if you remain within reach."

"The coppers 'll see that I don't; they'll keep me on to it."

There was something in her tone which struck his ear.

"In what way?"

"They've as good as said that I'd better look out what I was up to, because they were going to keep their eyes on me, so I've had to do a bunk as it is. Why, they gave me a room in the station, and locked me up in it; so when I got your message—"

"You did get it?"

"What do you think? There's ways of getting messages from your pals even when you are as good as quodded—and that Ike's a downy one. So, as the door was locked, and I didn't care to tell them that I wanted to take a little walk—they ask such questions, those coppers do—I got out of the window, slid down the slates, got on to the wall, and did a jump. Lucky it was dark, or I mightn't have found it easy. Before now they'll have found out I'm missing; they'll be wondering where I am; they'll be looking for me too."

"Do you think they will?"

"I'm sure. I know them coppers."

Mr Polhurston had returned to the fireplace. It was some moments before he spoke; he seemed to be seeking inspiration from the burning coals.

"This is worse than I supposed. You see how it is —if you are once my wife you can refuse with impunity to give evidence against the lad Foster; because, in any case, you will not be able to give evidence against me."

"How's that?"

"In a case of murder, a wife can't give evidence against her husband."

"Are you certain?"

"Absolutely. It is an elementary axiom of English law that in such a case husband and wife are one."

She got up from her seat trembling.

"Let me understand you. If I was your wife, couldn't I prove you gave me those things?"

"You could not. They could not put you into the box, and no statement you might make could be received as evidence. You could prove nothing; for any use you might be to the prosecution you might as well be dumb."

"Then, Gentleman, I—I wish I was your wife."

"So do I."

"If you was to marry me, you might give me the go-by directly after. You might do just as you pleased; you'd be free as ever. You'd be safe enough from me, because I wouldn't tell anyone you was my husband, except—in case."

"You would rather I gave you the go-by directly after?"

"You see, it's like this. Of course I should know that I could never be your proper wife; it'd only be, as I say, in case."

"Why shouldn't you be my proper wife?"

"Me! Me! Me!"

She held out her hands in front of her, wider and wider apart with each repetition of the pronoun. He mimicked her action.

"Yes! You! You! You!"

"Do I look as if I could? Do I sound as if I could? Ask yourself the question, that's all!"

"I am asking myself the question."

"Very well, then, isn't that enough? You're a gentleman—you've got lots of money—and, I'll be bound, heaps of friends of your own sort. You want a lady for a wife, and I shall never be a lady, not if I live to the end of the world."

" With the gallows looming at me through the air ? "

"Seems to me you needn't be afraid of no gallows, not if what you say is right, and you was to marry me—in case. I'm the only one who can prove anything against you—aren't I the only one ? "

" Then are you suggesting that I should commit bigamy ? "

" Well, I don't know that I should go so far as that."

" But—excuse my interrupting you, and forgive me if I misconstrue your meaning—if you're my wife, yet not my wife, am I to go without a wife, or, in search of a wife that is a wife—to become a bigamist? Before you answer, I should like to point out to you that, so far as I know myself, my instincts are not a bachelor's."

Her eyes twinkled, her cheeks flushed.

" Now, you're making game of me. Go on! Have your fling."

"One other question I would like to put to you. Don't you care for me ? "

" Course I do. What's that got to do with it ? Why, if I could only get you out of the mess by hanging instead of you I'm game straight off! My sakes! I only wish I had the chance ! "

She drew herself erect, her fists clenched; in her voice, her bearing, was sudden passion. As he looked at her, his face glowed.

" Then do you take it for granted that I don't care for you ? "

" Course you care for me. Haven't we seen enough of each other for me to know that ? Think I'm silly ? Only, things are different, now you've—come into your own."

" My own ! "

" Well, somebody else's own, then. It's the same thing, isn't it ? "

" Pollie, you're more of a philosopher than I imagined; and, I believe, of the very latest school."

" Chaff away ! I don't mind ! So long as I get you out of the mess I got you into, that's all I want —every bit ! "

" Come here."

" What for ? "

" I have listened, with exemplary patience, to your ideas upon the subject. Now I should like to tell you, in some detail, exactly what it is that I propose; and, while doing so, should prefer to have you conveniently close at hand. Come ! "

She went towards him.

" Closer ! "

Putting his arm about her, he drew her to him.

" Don't ! You "—there was a catching in her voice —" make me all of a shake."

" Do I ? We are even. Your near neighbourhood makes me tremble, which is odd though not unpleasant. Do you find it unpleasant to be—all of a shake ? "

" Don't ask me such questions ! Don't ! Go on with what you were going to talk about."

" I only wish to make you understand what a very comfortable thing it is to have you near."

" Don't talk like that ! Go on ! "

" Without the least suspicion of further delay I propose, as I have said, to marry you, with your permission, and have even gone so far as to give notice of my intention without first asking your permission."

" What do you mean ? "

" The nearest road to marriage is through the registrar's door. If it is your intention to marry by special licence, which, saving your presence, is my intention, the law requires that, between the giving notice and the marriage itself, there shall be, at least, one clear day. So, since time is of importance, I took the liberty, this afternoon, to go into a registrar's office, and to announce that, the day after to-morrow, I intended to marry you."

"Gentleman! You didn't!"

"Pollie, I did! Have you any just cause or impediment to allege why the marriage should not take place?" The girl was still; he felt her tremble like a leaf. "And since my ideas are different to yours, and I do not propose to give you the go-by, or to allow you to give me the go-by either, but to have and to hold you as my very real and very actual wife, I intend to go straight from the registrar's office to the Continent with you. Thence, by ways of which I know something, we will journey together to America. In that tolerably large country, I make no doubt that we shall be able to find some place in which you and I will be able to live without being worried by uncomfortable questions."

There was a pause; she had hidden her face on his shoulder.

"Are you—going to take me—with you—to America?"

"I am; with—I am more than half-inclined to say, even without—your leave, my lady."

"Oh—h!"

She gave a great sigh, as if her feelings were too much for words, then was assailed by sudden doubt.

"But—you'll lose your fortune."

"How so? It's not dependent on my place of habitation. If I choose, I can take it with me."

"I'm not a fit wife for you."

"That's your opinion; I'm sorry to say I'm of a different mind."

"You'll be ashamed of me."

"Very good, I'll be ashamed."

"You can give me the go-by whenever you like."

"Thank you; I'll keep that little fact continually in my memory."

Presently she went off on a fresh current of thought.

"But—I've no things—except these."

She looked down at herself, with doubtful eyes.

"Well, those are a good foundation on which to build, and mind you, in a case like this, a foundation's not to be despised. I tell you what—to-morrow we'll go shopping, you and I, and we'll buy everything that's necessary for the trousseau of a bride who intends to travel."

"Oh—h!" This was another great sigh. "But I sha'n't know what to get."

"Nor I; that's a matter of no importance."

"They'll laugh at us."

"Let them! We'll laugh back."

"You're laughing at me now—you're not to."

"I won't, you may depend on it."

"I don't know what things a lady ought to wear."

"Well, I'll give you a hint; expense is a point to be considered. The things which cost most money—those are the things which a lady ought to wear. When you've arrived at that conclusion you've got some distance."

"If I go into a shop and ask for things, they'll—make game of me."

"Then you sha'n't go into a shop and ask for things; I tell you what we'll do. We'll draw up a list, put everything on it, everything—down to the shoe-laces, and the ribbons, and the thirteen trunks to hold the entire collection! And we'll brandish it in the shopman's face and say, 'Produce everything which is on that list—at once!' And he'll produce it; and we shall look on, and say nothing; only smile."

"But—I sha'n't know what to put down on it.

"Nor I; so, if we both know nothing and put our heads together, and, as it were, pool what we don't know, we may find out.   At any rate, I propose we try."

They sat down at the table, or rather he sat, and she knelt on the floor at his side, and they tried.

The compilation of that list was unconscionably long in doing; and, by the time it was done, he was young again, and she—she was still a child.

CHAPTER XVI

A GLIMPSE OF SUMMER

IT was the strangest day, that which followed. The weather had changed. The glass had risen, the skies were clear, the sun was shining, there was no wind. Both were up betimes, especially considering the hour at which they had retired—with the list to blame—but Pollie was first.

When she woke up she wondered if she dreamed. Was this hotel bedroom an apartment in the palace of which she had read in penny numbers—the palace in which the Duke of Morbihan resided, or the Prince of Putney, or the latest scion of a line of countless earls? How soft was the bed on which she was lying! When one is unused to beds, one appreciates the pliancy of a palliasse, and the luxury of a mattress with decent springs! Where was she? She remembered. And to-morrow was her wedding-day!

She sat up in bed with a start.

To-morrow—her wedding-day! She had to say it over to herself before she could believe the thing was true. Her whole frame was in a flutter of excitement. And such a wedding-day it was to be!

She sprang on to the floor. There were her clothes, which he had bought her. Seating herself on the bedside, she examined them, inch by inch, stitch by

stitch.   What magnificent garments they were—a
lady's! And they were hers! And she was to have
more to-day. As she recalled some of the items on
that list, her pulses glowed. The thing was altogether too incredible.

She stole into the bathroom and had a bath, issuing
therefrom a picture of radiant health and gipsy
beauty. She dressed herself—a process which took
some time. She was not quite sure how some of the
garments ought to be worn — trying them on in
different ways; deciding, at last, with a mind all
dubitation.

The dress of dark blue serge, considering that it
had been bought at a venture, ready-made, was not
by any means ill-fitting. She admired the silken
linings, leaning strongly to the opinion that, for
their sake, it would be almost more in accordance
with the perfect order of things, to wear it inside out.
As she surveyed her reflection in the mirror she
decided that, on the whole, it did not become her ill.
If only her hair had been longer, and the ends more
even! How much time would it take for it to
grow?

In the next room she found there was a maid at
work. She crept out again, abashed. When she
ventured in again, the maid was gone. The fire
burned brightly; the table was laid for breakfast.
How splendid it all looked! Did ladies always live
amid surroundings of such splendour? She looked
at the things upon the table, anxious to learn their
uses, so that her ignorance might not shame her.
While she was turning a butter knife over and over,
wondering what its place in creation might chance to
be, the Gentleman, coming in, caught her with it in
her hand. She dropped it with a clatter, her cheeks
all flushed. Without a word, putting his hands upon
her shoulders, he drew her to him and kissed her;

whereupon she burst into tears. She had not thought that she was such a fool.

That shopping was a curiosity in procedure. Her disposition was towards remaining in the room all day, and not showing her face outside. She had, all at once, become afraid. But he would have none of it. He ordered a brougham from the hotel, declaring that in it she would be safe. And it seemed that she was; though by no means at her ease. Her shyness was continually to the front. She was afraid of him; much more fearful of all the world beside.

In every instance she needed persuasion before she could be induced to cross the threshold of a shop; all the time she was in, she was on thorns. That she was the observed of all observers was plain. While the fact occasioned him something like amusement, it caused her agony. She not only treated the assistants who sold the goods with most unusual respect and deference; she even stood in awe of the goods which they were selling, being with difficulty induced to handle them at all, as if she feared that a touch would work their ruin.

" Whatever shall I do with all those things ? " she asked, when they were back again in the carriage.

" Why, wear them, goose ! "

" Wear them ! " She looked at him with her big eyes open at their widest. " But I never shall be able to wear them all—never ! And there's most of them too good to be worn—at least by the likes of me."

" You'll change your mind about all that."

She drew a deep breath, as if she doubted it. Presently she said, in a whisper,—

" I'm afraid."

" So it would seem ; though it's a revelation to me to learn that you're a coward."

" I'm not—not in the ordinary way ; but this is different."

"How different?

He was observing her with curious glances. She gave another of her big sighs, which reminded him, in some queer way, of a child's sigh, half of rapture, half of wonder, on beholding for the first time the marvels of a pantomime.

"It's against nature."

"How is it against nature?"

"That I should be sitting in a carriage with you, dressed like a lady, and buying all those lovely things."

"Pollie, you're an idiot. Do you suppose that what you call 'gentlemen' only marry what you call 'ladies'?"

"I don't know nothing about that. I only know that I'm no fit wife for you—and you know it, too."

"Do you know how handsome you are?"

"Go on! Don't talk such stuff!"

"If I were to dress you as you might be dressed, and were then to show you on the stage of any theatre in London, you'd have, at any rate, all the men in the house at your feet."

"I don't want to show myself at the theatre. Let them do it as likes that sort of thing, I don't."

"I am not saying, or suggesting, that you do. I am only stating a fact."

"Look here—am I decent-looking? Tell me straight."

She turned towards him, sitting so that the light shining through both windows lit up her eager face.

"You are. We're now in Regent Street. I'm willing to lay odds that, though you walk from end to end of it, you'll not find a woman more decent-looking."

"Honest?"

"Honest. A woman's looks depend a great deal upon the way in which she dresses; you may take

my word for that. It's the same with a man. Am
not I better-looking now than I was—the night
before last?"

"Rather! I should think you were! You don't
look the same one bit."

Her frankness seemed to amuse him.

"If clothes makes such a change in me, what will
they do in you? When you see yourself in the
glass, decked out in some of the splendours which
we've been buying, you won't need me or any one
else to tell you if you're handsome; you'll fall in
love with yourself, my dear."

"Fall in love with myself! Sure to do that, I am.
But if I really am nice-looking it might make up for
it a bit, mightn't it?"

He knew what she was thinking of, and laughed.

"It might; I shouldn't be surprised if it did." He
changed the subject. "I don't feel like going back
to the hotel, especially in such weather. Pollie, have
you ever been to Richmond?"

She had not. She did not even seem to know
where Richmond was. So he took her, there and
then. They drove through Piccadilly and the Ham-
mersmith Road, over the Bridge, through Barnes and
Mortlake into the Park; she content for the most
part to sit perfectly still, as if the mere delight of
being driven was enjoyment enough for her.

He sat back in his corner, watching her, as she sat
back in hers, wondering, as he noted the expression
on her face, of what it was that she was thinking.
But he did not ask; he was content to wonder. For,
since the morning a glow, as of satisfaction, had been
mounting higher and higher in his veins; until now
it positively seemed as if there shone on the world,
this once again, the light which never was on sea or
land—youth's golden haze. It was odd.

At the bottom of the hill they got out and walked

up to the Richmond Gate. It is true that under foot, it was not particularly dry. But the snow had gone, and on the turf, that close-woven, springy turf, through which the water drains as if by magic, it was not so bad. And they had neither of them been accustomed to pay much attention to damp feet. Most of the way they went up like two children, hand in hand. No one seeing them would have supposed that, only a few hours before, they had been ragged, hungry, flying for their lives from the police, over the roofs, in the snow and wind at night.

The metamorphosis, outwardly, was complete. He bore himself as became a gentleman. And if the student of physiognomy, who saw his face, might have inclined strongly to the suspicion that he had lived, he would hardly have supposed that it was in any fashion unbecoming to a man of breeding, birth, and means. And she was such a handsome wench, and held herself so straight, and looked at the world so boldly out of her big, black, velvet eyes, that, though one felt that there was something *bizarre* about the way in which she moved and wore her clothes, yet one forgave it her because of the atmosphere of life and health and vigour which she carried with her as she went.

They stayed at the Star and Garter long enough to give the horse a bait, then, through the gathering darkness, drove back to the hotel.

"It's been a pleasant day."

She drew one of her big breaths before she answered, as if she wanted to get as much air as possible into her lungs.

"Pleasant! That's not the word. It's been my only day."

"And to-morrow we're to be married. Won't that be another of your only days?"

She did not answer. Within the vehicle it was

dark, so that he could not see her face; but, presently, by the sound he knew that she was crying.

"Pollie, what's the matter? You're showing a sudden fondness for tears."

"I'm afraid!" she said.

"Afraid? What of?"

Again she did not answer; and in the silence, over his own soul there crept a sense of fear.

At their journey's end, as they alighted, a policeman chanced to be standing at the door of the hotel. When they reached their sitting-room, Mr Polhurston found that the colour had gone from Pollie's cheeks. She was trembling.

"He knew me!"

She looked at him with eyes of frenzy.

"Who knew you?"

"That copper at the door."

"Nonsense!"

"He did—he looked at me—I saw him!"

"Well, even a policeman may look at a person without conscious recognition. I take it that you're not known to all the constables in town."

"But I've seen him before, and he's seen me. If he should come in here!" There came a tapping at the door. She clutched Mr Polhurston by the arm. "Don't let him come in, for Gawd's sake, don't!"

The door was opened, without waiting for permission. A waiter entered, to learn if they had any commands for dinner. If anything about the attitude of the pair—the woman clinging to the man as if for life, the man looking round with startled mien—struck him as singular, nothing in his demeanour showed it. Mr Polhurston issued the necessary instructions, the waiter went. Then Mr Polhurston turned on Pollie.

M

"My dear child, what is it which has come to you?
I do not remember you as a nervous subject, or
fanciful; certainly not as a coward."

Her alarm had not entirely abated, though shame
had partly taken its place.

"That was different; then I'd nothing to lose;
now, don't you see?"

He did see—perceiving her meaning as by a
lightning flash. When they were ragged, cold,
hungry, destitute, then all risks were none; the
transition from bad to worse could be, in any case,
so slight. Now to pass from the warm light in
which they were, to the outer, hideous, eternal darkness which threatened beyond, there was horror,
terror, in the thought alone.

It was this reflection which caused him to be
oppressed by an indefinable sense of dread, which
she did nothing to lessen. She still clung to his arm.

"Let's go away from here!"

He tried to laugh away her fears.

"What? At once? Before dinner? No, thank
you. Especially as, in any case, we shall be leaving
in the morning."

"Let's go to-night."

"Where to?"

"To where you said—America!"

"You're dreaming. Come, don't be silly.
America's not just over the way, as you seem to
suppose. In the morning it will be time enough
to start.

"It will be too late."

"Nonsense! You're over-strung. The excitement
has been too much for you. You shall have some
dinner, and then go straight to bed. A good night's
sleep will drive all the cobwebs out of your brain,
and in the morning you'll laugh at your own
fancies."

She rubbed her hand across her brow—as if her head was feverish.

"I wish you'd come to-night—I wish you would."

He led her to her bedroom, laughingly

"Be so good, young lady, as to go in there and prepare for eating. It's just possible that you may find that a sponge will freshen you—only don't be long. You heard me order dinner to be ready in half an hour, and here punctuality's observed."

The meal was scarcely a success. Neither seemed to have much appetite; there was little conversation. She hardly spoke at all. The pretence at light-heartedness with which he endeavoured to brighten the occasion was so signal a failure that very soon he gave it up, relapsing into silence too. Almost as soon as dinner was concluded, acting on his suggestion, she retired to her own room—if not to rest.

There was singularity in the fashion of her good-night. Kneeling by his chair, holding him close to her, she kissed him twice, thrice, and yet again. She had been, apparently, so cold and self-contained that the display of affection took him by surprise.

"I hope you won't allow marriage to put a period to this kind of thing?"

She did not reply to him directly, seeming to be following a train of thought of her own.

"You'll think of me?" she said.

"Sleeping and waking, the whole night long."

She eyed him earnestly, her glance conveying the impression that he had not said exactly what she would have wished him to. Her next words seemed to be pregnant with a meaning which was hardly on the surface.

"I'll think of you, and of to-day."

He exchanged with her look for look, with an air of seriousness in which there was more than a touch of whimsicality.

"I don't think we shall either of us forget to-day
—ever."

"Never!"

She got up and went to her bedroom.   And, as she
went, he heard her sob.

CHAPTER XVII

AT DEAD OF NIGHT

THE church clock struck two. The reverberation of the second stroke died away as she opened her bedroom door but an inch or two. She listened. All was still. She opened it a little wider, listening yet. There was not a sound. The lights in the sitting-room were out, but the fire still burned brightly in the grate. The flickering flames illuminated the apartment by fits and starts, now dying away and leaving all in shadow, now flashing into fitful glamour, so as to throw particular objects into prominent relief.

She drew her shawl about her closer. She had not realised that she was cold until she saw the fire. She hesitated; should she get some of its warmth into her veins before she went? It would be one more memory to carry away with her, her last warm at his fire. She went to it, leaning over and extending her hands. By the light of its flames one could see that she was no longer clad in the costume which she had worn throughout the day, with such pride in its glory; she was once more in her rags.

After a while, standing up, she turned her face towards the door of Blaise Polhurston's room. She passed the fingers of her right hand to and fro between her lips, as if in a paroxysm of nervous irresolution.

"I wouldn't wake him! I wouldn't wake him!"

She repeated this to herself, over and over again, as if striving at self-conviction. She moved quickly to the door.

"Gentleman! Gentleman!" she whispered. There was passion in her voice, as well as pathos, and a wild despair. "Good-bye! I'd never have been the kind of wife you ought to have, never! It's best like this—best for both of us. Gawd bless you! Are you dreaming of me?"

There was a break in her speech; a sob in her throat. Bending forward, she pressed her lips against the panel. And that same instant the door was opened from within. The handle was turned with a sudden click, the door drawn right back upon its hinges, and Blaise Polhurston came hastily striding into the room.

The thing was done so suddenly that the girl still had her lips pressed against the panels when, as it seemed, they were snatched away from her; she had no time to withdraw before Blaise Polhurston was pushing past.

"Gentleman!" she cried. Cowering, she shrank back against the wall. She expected him to turn upon her with reproaches; at least to demand an explanation of her presence there. But he did neither. It was a minute or more before she realised that she was unperceived, that he ignored her utterly. When the consciousness that this was so began to dawn upon her she was filled with sensations of mingled wonder and awe.

He was only partly dressed, having thrown a dressing-gown over his trousers and nightshirt. The dressing-gown, being unfastened, flapped open as he moved, seeming to throw his tall, thin figure into ominous relief. His hair was dishevelled. He carried no light, but, as he entered, a coal fell down in the

fire; the flames shot up, as if in greeting. He moved
to the centre, and paused, the girl momentarily expecting him to turn and discover her, wondering how
he could have passed her undetected.

Presently he began to speak—she perceiving with
increasing surprise that it was to himself he spoke,
apparently oblivious of the possibility of his being
overheard. There was an odd quality in his voice—
frenzy, agony, as if the words were being wrung from
him in clots of blood. The first word he uttered was
her name.

"Pollie! Pollie! Pollie!" three times over, in a
crescendo scale.

The girl, crouching in the corner, began to tremble.
What was coming? What was about to happen?
It was strange to hear him call to her, as it seemed,
from the depth of his heart, he remaining all the time
unconscious of her close propinquity.

"She's ill! How ill! Fallen down in the street,
and carried home. Home! What a home! Hunger,
weakness, exhaustion—that's what it is. Cold, ex-
posure, scanty clothing, no food—that's who's laid
her here. A doctor 'd say, give her plenty of nourish-
ing food; good wine—that's the medicine he'd pre-
scribe for a case like this. Ye gods!" He laughed
—a laugh which set his hearer shivering.

A flood of memory swept over her. She remembered how, during a spell of bitter weather, luck had
gone dead against them; how, practically, for days
together they had had to go without food—how, at
last, she had fallen in the street, and been borne,
senseless, to the garret which they called home—how
he had returned, and found her still unconscious.
Was this time being brought back to his nightmare-
haunted brain, so that asleep he was enacting his
dream before her eyes? So soon as she suspected
that this might be the case, her limbs seemed to

become paralysed; as if spellbound she continued to watch him.

She saw him, by the flickering firelight, drop upon one knee, and bend forward, as if stooping over some object which he perceived to be in front of him. Extending his left arm, he moved his hand gently to and fro, imitating the gesture with which one smooths a person's brow.

"Pollie! Pollie!" With what tenderness he pronounced the name! How her heart melted within her as she heard! "How cold you are! How cold—and how still! Pollie!" Again the clinging tenderness. The tears started to the listener's eyes. "Wake up—open your eyes—look at me, child! It's Gentleman! Don't lie so still—it's the first time I've asked you to look at me in vain!"

For a few seconds longer he continued the gentle movement of his hand. Then his arm dropped to his side, as with a gesture of despair.

"What shall I do? What shall I do? Starving! In this great storehouse of food!"

He stood up, remaining motionless, with head a little bowed, as if looking down at the silent figure which was at his side.

"Where shall I find the wherewithal with which to buy her food? I've begged—begged!" He laughed—again that uncomfortable laugh which had set her shivering. "I've not found the profits which they say the trade brings in. Perhaps that's because I am unpractised. There's no time to gain proficiency. I must get food for her to-night; and—for me. But how? How? With nothing how shall I get something? How? How?"

She saw him shudder. He threw up his arms, as if, in so doing, he strove to throw from off his mind some thought of horror.

"What devils hunger makes of us! What devils! Or else it lays us there!"

He pointed to where he perceived the silent figure on the mattress at his feet. Turning, he began to move hurriedly, jerkily, to and fro, as if he was endeavouring, under pressure, to resolve a problem in his bewildered brain. Words came from him in disconnected phrases.

"There is not time. I could not do it if there were. An hour or two, it may be too late. Money —money—at once, or not at all. They must bury, though, it seems, they are not forced to keep alive; that's how it stands."

He paused in his nervous movements as if to consider. A name came from his lips as if unconsciously.

"Shapcott! Shapcott!" Then, as if waking all at once to comprehension of the thing which he had said, he extended his arms in a burst of sudden passion. "Curse him! Curse him! May he be trebly damned!"

Again he dropped beside the imaginary pallet, this time on both his knees.

"Pollie! Pollie!" There were tears in his voice, which increased the streams which were pouring down the listener's cheeks. "Don't lie there so still— don't, don't! Open your eyes—look up at me—speak to me, Pollie, speak."

But, apparently, his appeal remained unheeded. On his face, as revealed by the firelight, was an agony of supplication.

"Pollie! Pollie! Pollie!" he murmured, over and over again.

Then he sprang back to his feet, with a sudden access of determination.

"She shall not die, that is certain; not if any action of mine can keep her alive. There are ways

and means of getting money—always ways and
means. I have heard it said a hundred thousand
times—I know it of my own knowledge. Why
should I stop to consider what things are lawful
when there's death to fight? Against that foe all
things are lawful, God knows! Death comes quick;
he does not wait, cap in hand, for you to stand considering. You must beat him with any weapon—
yes, at once!—now!—or you'll be beaten; and
that defeat there's no retrieving. God knows! I'll
go at once to Shapcott, and I'll beg of him for
her life."

He hesitated, as if he had still not succeeded in
bringing himself quite to the sticking-point. Then,
once more knelt down.

"Pollie, it's for you. God help us both!"

He bent his head right over, as if stooping to kiss
the face of the silent figure which was lying on the
pallet. And, indeed, the listener seemed to feel his
lips touch hers.

When again he rose to his feet, his words and
manner suggested that the scene in his vision had
changed, as, in our dreams, scenes do change, with
kaleidoscopic suddenness. His attitude seemed a
rigidity of horror which it was unpleasant even to
contemplate. He stood so that the fire shone on his
features; nothing in the expression of his face, or in
his bearing, was hidden from the girl, staring, all
trembling, from her corner.

He looked like a man might look who sees a ghost;
as if he were compelled to glare, willy-nilly, at some
object of horror, the mere sight of which froze the
blood in his veins. As Pollie watched, something of
the horror which held him as in a vice passed from
him to her, so that she became an involuntary sharer
in his nightmare agony.

He was staring downward at the floor, with parted

lips, wide-open eyes, as if there were something there to which his glance was riveted.   He continued to stare so long, so fixedly, without moving, or, apparently, drawing breath, that his strained pose began almost to suggest paralysis—some dreadful form of tetanus which held him rigid.

It was a relief when, at last, a tremor passing over him, with a long, deep-drawn sigh, he seemed to return to life. For some seconds he swayed to and fro, as if he were about to fall. Then, all at once, drawing himself quite straight, he looked round the room, with unconscious, glassy eyes. Then back went his glance to the floor. Something of his former dreadful look returned. He muttered to himself a name, hoarsely.

"Shapcott! Shapcott!"

As if startled at the sound of his own voice, again he looked about him, with the same unseeing eyes. Once more they returned to the something which he seemed to see lying at his feet. He slightly inclined his head.

"Is it you? What—what are you doing there?" He paused, as if for a reply. "Shapcott!" Another pause. "What—what are you doing there? Why don't you speak to me? Shapcott!"

For the third time his oblivious gaze passed round the room. "I thought I heard a noise. Yet it all seems still. How still! I wonder what's made the world so still! Shapcott! Why the devil won't he speak to me? Why the devil? I suppose it is Shapcott. I can't see him with his face turned down. Shapcott!"

He reached out his foot, as if touching something with his toe. Then put his hand up to his throat, as if labouring for breath. His voice had become curiously hoarse.

"This won't do. I musn't stand here, doing

nothing, like a fool; I must do something; I must find out what's the matter with Shapcott.

He lowered himself on to one knee, gingerly extended his right hand, and started back with a cry.

"He moved! I thought he moved!"

He was trembling as with palsy; the sweat stood in beads upon his brow.

"I was mistaken. It was a delusion—one of those delusions which come to people whose nerves are over-strung. Where was I reading about tricks which can be played with the eyes? Besides, why shouldn't he move? I'm a fool. Shapcott!"

Again he advanced his hand, and again withdrew.

"I seem to be afraid to touch him. Why should I be afraid? It's absurd. It's only Shapcott. I'm all run down, that's what's the mischief. I wonder what's in that bottle?"

He got up and, taking two or three steps, went through the action of taking up a bottle, and putting it to his nose.

"It smells like whisky. Shapcott, may I have some of your whisky? There must be something wrong, or he would certainly say no. I shall interpret your silence as implying consent. Here's to your health, Shapcott, and your happiness, wherever you are." He made as if he raised the bottle to his lips, and drank from it—a big drink. "Why, he's on the floor!"

He laughed; seldom did laughter seem to be a thing so little to be desired. He returned to where he had previously stood.

"Come, let's make an end of this. To see if anything ails Shapcott really—if so, what? Perhaps he's drunk."

Once more that undesirable laughter. Again he knelt, this time on both knees.

"Now, Shapcott, rouse yourself; don't lie there

like a log. If it's a trick you're playing it's in worse taste even than some of the other tricks you've played on me. What's the matter with you, man?"

Again his hand went out, not on this occasion to be immediately withdrawn. He seemed to be touching something with unwilling fingers.

"Shapcott! His arm is limp; there seems to be no pulse. How still he is!—God in Heaven! He's dead!"

The word burst from his lips with a cry in which anguish, horror, and surprise were mingled. The discovery—it seemed to be a discovery which he had made — appeared to overwhelm him. The stupefaction which was mirrored on his countenance was the apparent outcome of a dozen different emotions. At that moment he looked like a man who was unable to believe the evidence of his own senses; and feared to.

"Dead!"

He covered his face with his hands, trembling; removing them, he looked about him as if frightened by the shock of some unlooked-for, some incredible event. He clambered to his feet, awkwardly, as if not wholly master of his limbs. His fearful glances wandered hither and thither.

"How can it have happened?" His own question went unanswered. Recollection seemed to come to him. "Perhaps Pollie's—dead — of hunger. God help us all!"

Again he put his hands up to his face, shivering. Presently, he seemed to make an effort to better play the man.

"Well, since what's done is done, there's no undoing it; that, I take it, is an axiom. How it came to be done is hardly matter, for my present consideration. Am I judge and jury? Or is it my office to sit upon the corpse, or to place a prisoner at the

bar? Go to! these things belong to others. I came
for money, and what behoves me is to go with it.
Perhaps there's money in his pockets—or in a drawer.

"But, no doubt the drawer is locked; and for his
pockets, I'll none of them; they're much too close.
Even yet I'm not brought far enough to pilfer from
a corpse—of whoever's making. If I can't have
money, I'll have its worth. There! there's gold in
them, and, maybe, life for Pollie."

Taking a step or two to one side, he made as if he
picked up something—some small object with either
hand; then, with a sudden start he looked round.
His voice was raised, eager words broke from his lips.

"Who's that? It's someone outside the window!
I heard him move! I saw him go! Perhaps he did
it—that's how it happened. Stop there, stop! Who
are you running away with the stain of blood guilti-
ness upon your soul? Stop! stop! stop!"

He rushed to where the curtains shrouded a window,
and began tearing at the hangings with impetuous
haste, as if he strove, with might and main, to get
at something which was beyond. Pollie, standing
up in the corner, ran to him, and put her arms about
him.

"Gentleman! Gentleman! Don't! don't! It's
me—it's Pollie!"

She was crying as if her heart would break.

At the sound of her voice, at the touch of her
arms, he relaxed his violent assault upon the curtains.
He was still. Then he began to shiver, till Pollie
thought that he would never stop. She held him
closely to her, as if she hoped to soothe and calm
him by the strenuosity of her embrace.

"S'sh! s'sh! It's me! it's Pollie! Don't you
know me, Gentleman?"

By degrees his shivering grew less. Presently he
turned and looked at her, with consciousness.

"Pollie? Thank God, it's you. I think—that I've been dreaming; but I'm glad—I've woke—to find you here. You mustn't leave me—ever."

"No, Gentleman, I won't—never—as long as I live!"

The flame in the fireplace died down, the room was all in darkness.

"Pollie! Pollie! Pollie!"

He could be heard repeating the name, a little faintly, in the sudden blackness, as if the sound of it brought solace to his soul.

She held him in her arms, supporting him as if he had been some weakling child, raining kisses on his face, while her scalding tears dropped on his haggard cheeks, embracing him, striving to comfort him, crying out her heart, all in a breath.

CHAPTER XVIII

ABOUT the meeting in the morning there was an atmosphere of strangeness; as if the relation in which each stood towards the other was a new one, to which they had yet to become accustomed.

Neither put in an unduly early appearance in the sitting-room. Pollie was first. Once more the rags had disappeared; again she was in the blue serge dress of yesterday. A certain air of uncouthness which had been in her bearing had become less marked. There was about her manner, her movements, her speech, a flavour of responsibility, of authority, which improved her not a little. She had become less self-conscious; her partial forgetfulness of self seemed to have equipped her with a peculiar dignity.

When Mr Polhurston entered, she was standing before the fire, whose flames had lit the scene the night before, staring down at the blazing coals as if, in them, she strove to read something of her destiny. He advanced to her with both hands held out.

"Pollie, this is our wedding day. I am very glad. I hope you are not sorry?"

Her blood showed more clearly through her cheeks.

"You know I'm not."

He kissed her, first on either cheek, then on the

192

lips. She continued in his arms for a moment, silent, trembling; the disposition towards the tremulous appearing to have passed from him to her.

When they sat down to breakfast, no direct reference was made to the events of the night, but she, peeping at him out of the corners of her eyes, perceived that they had left their traces on his face. Despite the punctilious perfection of his attire, designed, no doubt, to give him the appearance of buoyancy and ease, he looked worn, nervous and haggard. The air of whimsy, which was his predominant characteristic, for once in a way sat on him ill. His efforts to be whimsical were laboured.

"Treasures all arrived? Is there much packing to be done? If so, you must summon the assistance of the maid."

"All the things have come. I was awake early, and packed up everything."

"Well done, for the time is short. It's half-past eight, and the marriage is at ten. I'm told that, on these occasions, when the registrar is the officiating priest, and there are no parsons, no bridesmaids, no singing boys, for whose several and collective absence may the saints be thanked—the ceremony's brief and to the point. Mr Registrar asks you if you are willing to become my wife—I hope you'll not answer no."

"You know I won't."

"He then asks me if I am wishful to become your husband, which in very truth and deed I am; then, when we both have answered, the thing is done. Which will be just as well, for we have to be at Charing Cross at eleven, to catch the mail for Paris."

"For Paris?"

"Where we'll stay long enough to just look round, before making up our minds to go on further."

"I wish we were there!"

N

"So do I." He paused; then added, "Because then you'll be my wife, my dear."

She said nothing. It was to be noticed that all his little attempts at gallantry she allowed to go unheeded—at least, so far as an answer was concerned. And, indeed, they seemed to have an exactly contrary effect to that which he intended. Instead of lightening her spirits or amusing her, they seemed to add to the burden of depression which, quite obviously, was on her. Scarcely had she commenced the meal than she had finished; she had no appetite for food.

"Let's make haste! Let's get away from here as quick as we can!"

He, too, rose from his chair.

"Agreed! I'm as willing and as anxious to become one as you by any possibility can be—and find your impatience extremely flattering."

Her reply was candid.

"It isn't that. I want to get to Paris as fast as we can—while we're safe."

He looked at her with a smile. For the first time, in his tone there was a touch of genuine whimsicality.

"My dear Pollie, I do assure you that you need have no fears upon that score. You'll get to Paris in perfect safety; that is, if you esteem yourself safe with me."

She eyed him askance, as if not altogether clear as to his meaning; then ignored it.

"How soon can we get away from this place?"

"At once. I've only to call for the bill and pay it, which, as I happen to have some money, I trust won't take me long, and we'll be off. I, also, have done my packing; we need only arrange for our belongings to meet us at Charing Cross, then hurrah for the registrar and the Paris mail."

Again his attempt at joviality, and at the display of some sort of sentiment suited to the occasion, went disregarded. She was practical—only.

"Let's hurry! I'll go and put my things on now. I'll be ready in a minute. I sha'n't feel right till we're safe out of this."

She went into her bedroom. So soon as she was in, and had closed the door, she leaned against the wall, clasping her hands in front of her. Apprehension was written large upon her face; what she feared were coming events which cast their shadows before. Her anxiety lest, at this last moment, the cup—that undreamed-of cup of happiness which she was raising to her lips, and of which she had already, in anticipation, tasted—should be struck from her hand, and dashed in shattered fragments which never could be mended, to the ground; this anxiety made of her an unreasoning coward.

"If now—now—them coppers were to pinch me!"

The thought, acting on her as does despair on the hunted stag which is driven to bay, filled her with the sudden frenzy in a hopeless rage.

"If they do, I'll kill them—kill them!"

At that moment she looked as if she could, and would. There was a devil in her eyes.

The business of paying, of giving directions for the despatch of their luggage, was not a lengthy one. Shortly they were in a hansom, driving to be married. When they started, under cover of the apron, Mr Polhurston took her hand in his, saying, with a laugh,—

"This is our wedding journey."

But she, on her passage to the cab, had looked about her furtively, eyeing, with fearful glances, the people in the hall, the loungers on the steps, seeing in every face an enemy's, dreading danger on every side. And now, though the vehicle bore them rapidly on, her fears were still on the alert. She sat as far back as she could, as if she sought concealment, while her glances passed from side to side of the street, on the keen look-out for lurking foes.

"Why didn't you have a four-wheeler? We can be seen so plainly in this."

At this he laughed again, a little hollowly.

"Why should we not be seen? Are you ashamed of being about to be married? Or are you of opinion that a bride should court seclusion?"

"I expect that by now those coppers are looking for me high and low. I know they are."

"I fancy you overrate your importance, even to them. You are a free agent. They have no right to interfere with your comings and your goings."

"Right don't trouble them much, never you fear. I know them coppers. They'll have their knife in me because I have given them the slip; they won't think no trouble wasted so long as they can lay their hands on me again. And if they was to cop me now—"

She stopped, because of a choking something in her throat.

"You goose! I have a better acquaintance with the law than you. In the first place, they have not the faintest inkling of your whereabouts; dressed as you are I doubt if they would recognise you even if they saw you. And in any case you are only a witness. The English law does not regard a witness as a criminal. What they have done to you already they have done at their own serious peril. Certainly no policeman would dare to lay detaining hands upon you while I am here—be sure of that. Take heart, be yourself. Don't go to your wedding as if it were your funeral; think of what a poor compliment you are paying me."

But she would not be heartened, although he persevered. She pressed his hand and she tried to smile, but in the pressure was a tremour and in the smile a tear. And, somehow, his efforts to inspire her with the gaiety which he feigned to feel himself hardly rang sincere. One felt that his own mood was not entirely a festive one. It was a sober drive, theirs, to the wedding.

They turned into the street in which was the office of the registrar. Blaise Polhurston drew her attention to the fact with what was meant to be an air of triumph.

"See! Who is right, and who was wrong? Don't you hear the wedding-bells clashing in the air? There's the church, and within is the parson—and there's this advantage about our parson, that he'll but take a minute to rivet the bonds, which are to make us one. Come—here we are. Now, child, do wear a happier face; don't induce Mr Registrar to have suspicions that you're being dragged into this match against your own sweet will—that it's a marriage on which the sun's not shining!"

The only answer which she could be induced to make was,—

"I hope he will be quick, so that we sha'n't miss the train. I shall be all right when once we've started."

He pulled a face, as if he found in her speech a flavour of an acrid humour. The cab drew up. Hat in hand, he helped her to alight, whispering to her as she came,—

"Here's the church. My wife'll come out of the door through which you'll go in."

Another cab dashed up from behind. Two men sprang out. One of them touched the girl upon the shoulder.

"I want you, Pollie Hills."

His tone was curt, business-like. She turned and

stared at him, without a word. Mr Polhurston
interposed.

"What is the meaning of this? How dare you lay
your hand upon that lady?

The man replied, good-humouredly,—

"That's all right. She knows what it's all about.
She won't make a fuss. I've got a warrant in my
pocket; if she likes I'll read it to her when I get
her—she knows where. I don't know what little
game she was going to have with you." He glanced
at the registrar's office, in front of which they were
standing, and smiled. "But whatever it was, you
can take it from me that you're well out of it."

The two men, each having hold of one of the girl's
arms, conducted her into the cab. Mr Polhurston
made as if to follow them. He tried to speak. A
spasm, passing all over him, seemed, on a sudden, to
twist his limbs all out of shape. His face went an
ashen grey. He dropped to the ground like a log;
his hat falling off, his uncovered head struck against
the pavement as he fell.

CHAPTER XIX

TWO GIRLS

BLAISE POLHURSTON was going home. It was a curious sensation, as, in these latter days, most of his sensations were; he had them in abundance, and, about most of them, was an element of the bizarre. Leaning back in his corner, he laid his book on his knee, and he smiled; and the girl in the other corner smiled too.

He had suspected her of it before. Ever since they had left Swindon he had been laying down his book to smile; more than once he had been disposed to think that his amusement was amusing her. Now, there was not the slightest doubt of it. The smile was of a quality and quantity, which was not to be denied, even though she looked out of her own window with the air of one who took a thrilling interest in the landscape through which they flew.

She was a pretty girl, amazingly pretty. And dressed to perfection — or so it seemed to him. Laughing eyes looked out from under a picture hat; pouting lips, in some inscrutable fashion, suggested a little child. He told himself that she was a child—no more—whatever her years might be. And Mr Polhurston set them down, there and then, as seventeen.

It was plain that she knew he had caught her smiling; whence the air of sudden dignity, this turning of her face away from him, this intense admiration of the flying fields. But Mr Polhurston found her, all at once, of so attractive a personality, that he resolved to try if her sense of dignity was proof against the instinct for sociability which he felt persuaded she possessed.

"Going far?"

"I'm going to Cornwall."

As the word came from her lips, it seemed to strike a chord which was hidden somewhere in his breast, and which it set vibrating; and that, although the speaker's tone was cold, distant, full of an infinite reserve. He persevered.

"Do you know Cornwall?"

"I ought to. It's my home. I've lived there most of my life."

Still the voice was chilly, haughty, obviously intended to keep presumptuous advances at a distance. Yet the questioner went on, possibly feeling that a bearing of such extreme severity was hardly in keeping with an appearance of so much benignity.

"Do you know a place called Polhurston?"

"Polhurston?" As she repeated the word, ceasing to examine the landscape, she turned towards him, and stared—it was borne in upon him, more clearly than ever, what an exceedingly pretty girl she really was. "I know it very well."

"Is that so? Then possibly, you know some of the people thereabouts."

"Some of them."

"Perhaps"—he turned over the leaves of his book as if he wished her to understand that he was asking but the most casual question—"you know the old lady?"

"The old lady? You mean Mrs Polhurston? You wouldn't dare to speak of her as an old lady if she were in hearing."

"One would hardly discuss any lady's age if she herself were an auditor, would one?   But, still, isn't she well advanced in years?"

"I don't think that she's of that opinion."

"But doesn't she show any of the decrepitude of age?"

"I'm not so sure of that; but I do know that she wouldn't like you to hint at anything of the kind?"

"Indeed!   A lady of some strength of character?"

"Pretty well."

The girl's eyes were twinkling.

"Disposed to have her own way?"

"Why shouldn't she—in her own house—among her own people?"

"Precisely.   Why shouldn't she?"   Mr Polhurston's tone was dry.   "She lives alone?"

"No."

"Who lives with her?"

"Mamma and I."

In the utterance of the words was such a ring of laughter that that alone would have caused Mr Polhurston to start, even if he had not been taken aback by the statement which they conveyed.

"Mamma and you?   I beg your pardon; I had no idea."

"How should you have?   There's no harm done, only it seemed funny that you should be asking me such questions."

"Then you are—?"

"I'm Dolly Hamilton; my mother is Mrs Polhurston's daughter."

"Your mother is Mrs Polhurston's daughter? Then—you're her grandchild?"

"I am.   Does it seem so incredible?   How amazed you look."

"I had no idea she had a grandchild of—"

He hesitated; she urged him to continue.

" Of what ? "

" Of—such ripened years."

She laughed—the clear, ringing laughter of a happy child. He thought that, for many a day, he had not heard so musical a sound.

"Ripened years is good; I'm afraid she doesn't think they're ripened. I believe that in her heart of hearts she considers me what she'd call a chit—she's fond of those old-fashioned words. Of course, I'm not. But I find it difficult to persuade elderly people that you are grown up even when you are."

" Is that so? I was not aware of it."

"It's a fact. Look at me." He was looking, with a sense of unusual enjoyment. "I'm a woman, a full-grown woman—anyone can see that I am; besides, think of my age—positively ancient! But, do you know, I believe that mamma thinks, and I'm certain that grandmamma is completely convinced, that I am nothing but a child."

"Can it be possible? It seems to be eccentric behaviour upon their part."

She shot at him a quizzical glance.

"Now you're laughing at me—exactly as mamma does—and you a perfect stranger! You have been laughing all the time; I do hope it hasn't been at me."

"On my honour, no."

" To be laughed at is so trying to one's sense of what is due to one."

" So I should imagine."

" You've no idea how much at times I have to suffer. I don't parade it before the world, but I feel it all the same."

"That I can easily believe."

"Do you know Mrs Polhurston? You talk as if you did."

" I used to, once upon a time. It's many years since we met."

" Perhaps you also knew my mother ? "

" I fancy that I did."

Again there was that dryness in his tone. It seemed to escape his listener's ears. What also, apparently, escaped them both was the resemblance, indefinable, yet sufficiently real, which existed between them. It was most marked about the eyes. The whimsical light, which continually flashed in his found in hers the oddest echo. A casual observer, on the supposition that they were strangers, would again and again have found the coincident expressions not a little startling.

Now that her tongue was once set wagging, it continued. There was about her chatter a quality which was pleasant to his senses. It was light, graceful, irresponsible ; titillating his ears with a reminiscence of something which he had known, and loved, and lost, ah ! long ago. Though scarcely wise or witty, it was yet not foolish. She passed from theme to theme taking, as it were, a sip from each, and tossing it, like a bubble in the air, so that it was shot with sprays of colour. And since she was content to do the talking, and he to listen, they were good company.

At Exeter, a friend came into the carriage, another girl whose advent filled her with as much apparent pleasure and amazement as if she had been some good fairy fresh fallen from the skies. The new-comer was by two or three years the elder ; not so pretty, though pretty enough, of a larger build, and quite as talkative. They chattered one against the other. Mr Polhurston, unused these many years to the gabble of girls, in the din of their gay nonsense, found himself growing distinctly younger.

" Why, May Gifford ! " exclaimed Miss Dollie Hamilton, with a superfluity of italics, " who would have thought of seeing you ? Where have you sprung from ?   What do you mean by jumping into the train like this without the slightest notice ?"

"I suppose I can jump into the train as well as you, can't I ?   It's not your train."

"My dear, it is my train—my very own train! But that's a detail. The point is that you ought to let a creature know that you are coming, especially when a creature's thinking that you're at the other end of the world."

"I didn't expect to find you here. I thought you were at the North Pole."

"I have been—but I've returned. I'm returning now."

"You yourself told me you weren't coming back for ages, at any rate till the end of the summer."

"I never meant to. All my precious plans have been upset—tumbled topsy-turvy—every one of them ! A hundred things have happened since then— thousands. Haven't you heard ?"

"How should I ? My mind is occupied with serious things; I don't concern myself with trifles."

"Thank you. You civil thing ! Now I won't tell you."

Dollie Hamilton leaned back in her corner, with an air of offence.

"Don't !" May Gifford, leaning forward, tapped her friend's knees with her gloved fingers. "Do! do tell me ! "

Miss Hamilton unbent a little.

"Haven't you really heard ?"

"How should I hear! Don't be provoking. Haven't I been in an out-of-the-way corner, at the other end of nowhere, where all day long, morning, noon, and night, and in my dreams, I've heard of nothing else but golf—twosomes and foursomes, hazards and bunkers, lies and greens, puts and drives, irons and niblicks ! My dear, if during the next six months my best friend so much as mentions the word golf, I'll cast her off without a pang. I hate the game—can't play a bit. Every time I do play it takes me longer to get round than it did the time before; I don't know how it is, but so it is. It's hideous!"

"Then I'll tell you what; I'll play you a match next week."

"How many will you expect me to give you?"

"Give me! You conceited person! I'll play you level for a box of gloves."

"My dear child, you don't know what you're talking about. I have been doing some real playing, and I have been living with real players. Why, I can drive—it's either seventy yards or seventy feet, I'm not sure which, but I know it's one or the other."

"Driving, May, is nothing; putting is the art. You should see my play on the greens; I've been complimented on it by all sorts of people, really!"

Miss Gifford held up her forefinger.

"Look here, I've been listening to that sort of thing till my brain's become addled; if you say another word about it I'll throw you out of the window—now, you know. Tell me this wonderful news. What has made you change your plans?"

"He's been found."

"Has he? Where? On the roof? Or in the dustbin?"

Dollie shook her head solemnly.

"You think you're funny, but you're not—you're only silly."

"You certainly are not only silly, you're heaps of things besides, all of them worse. But what's been lost. That bandy-legged pug of yours?"

"Bandy-legged? Do you think that's a defect? Don't you know that all really good pugs are bandy-legged?"

"I don't care if they're no-legged. What's been lost, and what's been found, which has induced you to disorganise the whole course of Nature by returning before the time?"

"Blaise Polhurston!"

"No! You don't mean that!"

"I knew you'd say I didn't mean that. I was sure of it. It's just the sort of thing you would say."

"My dear girl, don't be so frightfully superior. Do you know that, at times, you're—almost—a prig?"

"A prig? Me? of all persons! You grow more civil with advancing years."

"You're not a prig—you're not! You're—I don't know what—but you are. Do you really mean that that black bogey's been unearthed at last?"

"Please don't call my uncle a black bogey."

"What kind of person do you yourself suppose he is?"

"Between ourselves, horrid. He must be, to behave as he has done—it stands to reason. Don't you think so?"

"Of course! Where was he found—in a dustbin?"

"I shouldn't be surprised if it were in something of the kind. But I know nothing—absolutely. I only know what I've been told, and that's about so much."

With the finger of one glove she marked off on the other glove a space of about a line in breadth—or less.

"They don't keep you thoroughly well posted up, then?"

"They don't. I believe they think I'm too young, or too feather-headed, or something. It's most ridiculous."

"Most! I suppose he hasn't turned up with a wife and about thirteen children?"

"Gracious, May! How horrid you are!"

"It's a way uncles of his sort have. I have a cousin who married a black lady in the Sandwich Isles; he left her there when he came home. She followed him, and wanted him to find posts in the army, navy, church, and civil service for about twenty little blackamoors. Nice for the rest of us, who like our cousins white."

"I hope my uncle won't be quite like that."

"We'll hope it as hard as ever we can. I suppose Mrs Polhurston is in the seventh heaven?"

"Higher than that, at least so I gather from what mother says. He is coming home. I don't know exactly when, but I've been ordered to return, so that I may be ready when he does come. He takes his time, but I'm not allowed to take mine, and I expect there'll be addresses, and triumphal arches, and brass bands and all the fun of the fair, and the county's to be invited to make a fuss, and he's to be asked to stand for the division; so I suppose he'll go up to Westminster, with flying colours, as a representative of all that a Member of Parliament ought not to be."

"That's why it's best to be a man for some things. The worse you behave the better you are treated. While we—oh, dear!"

"We're expected to be angels."

"Which, fortunately for ourselves, we mostly are. Of course, my dear, I'm speaking for myself; in such a matter I shouldn't dream of speaking for you. I hope I shall be allowed to see our uncle when he comes."

"I shouldn't be surprised; he's a bird of your feather."

"Dolly! That's rude! But I'll pass it by, because, as a matter of fact, I have a penchant for examining the features of regular villains. What sort of person would you suppose he is to look at?"

"That's his portrait which hangs in the hall."

" Yes, but that was taken when he was still a boy, before the brand of villainy was on his brow,"

" I don't suppose he's much altered. We Polhurstons have always been famous for our good looks, and for our knack of keeping them."

Miss Gifford held up her hands.

" Was ever such conceit! What a dreadful thing it must be for a family to cherish such a superstition as that, when only the superstition itself remains! Because you must find it difficult, at times, to induce people to believe that you are good-looking, merely because, according to the family legend, you ought to be. Sometimes people will credit the evidence of their own senses. Haven't you found it so?"

" We haven't all got twenty blackamoor cousins as yet."

" You may well say 'as yet.' In that case the worst is known. In the case of our Uncle Blaise, who knows what unimaginable crimes may be brought before the light of day, even to twice twenty blackamoors! But I'll forgive you for all the cruel things you've said, and all the cruel things you've thunk, and I'll tell you what. When you introduce me to our Uncle Blaise, if you like I'll kiss him for your sake!"

" I shouldn't be the least bit surprised if, before very long, you want to kiss him for his own."

" Dolly!"

When the train stopped at Plymouth the gentleman in the corner got out. The two girls began to discuss him the instant his back was turned. Said Miss Gifford,—

" Do you know him? He bowed and said good-bye as if he knew you."

" I was talking to him before you came in; he talks of having known grandmamma and mother too, once

upon a time. He seemed an inoffensive sort of person
—one of those colourless elderly men, the monotony
of whose lives has been interrupted by nothing more
remarkable than the vicissitudes of their annual
income."

"Sort of dry-as-dust; I know. But you can never
tell. An attitude of external reticence sometimes
hides consuming fires within."

"That's twaddle—penny noveletty."

"He hadn't bad eyes—they reminded me of you."

"Of me!"

"I don't mean to say that you haven't bad eyes,
because, of course, you have. But, every now and
then, when he looked round, there was something in
his eyes which I've seen over and over again in yours
—it's in them now."

"What an imaginative person you are!"

"And what an unobservant one you are. I'm persuaded that you spend more than half your time
before the looking-glass, so that one would think that
you would at least know what sort of eyes you really
have—but it seems you don't."

"Do you know that I'm beginning to suspect that
you're fond of making personal remarks which are
not in the best taste!"

"Are you? Think of that now! How precise
we're getting. With advancing years are you really
becoming proper? Well, there's room for improve-
ment in that direction—heaps. Do you never make
personal remarks, my dear?"

"Oh, yes, now and then, when the occasion requires
them—but then they are always in good taste."

"The conceit of her!"

"To know oneself is not to be conceited; it is what
we are all of us enjoined to do."

"And you think you know yourself? Dear, dear!
'Where ignorance is bliss!'"

o

Once more Miss Gifford raised her hands and the train travelled into Cornwall.

As Mr Blaise Polhurston was being driven through the Plymouth streets, his lips continued to formulate two words,—

"Uncle Blaise! Uncle Blaise!" They seemed to tickle him; for, after about the twentieth repetition, he smiled.

"How odd it sounds—her Uncle Blaise!"

CHAPTER XX

BLAISE POLHURSTON GOES HOME

It was dark—darker than he had expected. It confused him. The face of the country was changed—landmarks had been removed, objects with which he had been familiar, on whose continued presence he had confidently counted, had vanished; others, of which he knew nothing, had taken their place. Nothing was as it used to be. He had not counted on finding himself so entire a stranger in a place with every inch of which he had supposed himself to be intimately acquainted. The passage of the years had left its impress even on that remote corner of the world. He was inclined to be sorry that he had set out on the adventure.

And yet he had felt himself incapable of running the risk of encountering such a reception as that suggested by Miss Dolly Hamilton. Triumphal arches, brass bands, to welcome him home? That might not be. He would alight at the adjoining station towards the evening, and would walk through the shadows, across the intervening stretch of country, home. He would arrive there, unheralded, unannounced, with as little fuss or ceremony as if he had quitted it but an hour before. It would be best like that.

But he had not grasped the situation as it now presented itself to him. The night was unusually dark,

the station but a strip of platform which served a
scattered hamlet. With it he left behind all visible
signs of human existence. It was some six or seven
miles across country to Polhurston, through lanes,
over a strip of woodland, through the home woods.
That was as he remembered it. But he had not gone
very far before he discovered that, at least in this
light, his memory played him false. A gate against
which he suddenly came into contact, brought him to
a standstill, to discover that he had lost his bearings.
He had strayed off the proper road; the lane into
which he had struggled was, apparently, an *impasse*
—unless they had put a gate up since his time. He
retraced his steps, to presently plunge into what
seemed to be a labyrinth of twisting lanes.

He stopped. He had a compass on his chain. He
examined it with the aid of a match. So far as he
could judge, the lane in which he was, was leading
him in a wrong direction. And yet he could not be
sure. A few yards further it might turn—set him
straight upon his way. He tried to think where he
might be. In vain. There was nothing to lend
assistance to his eyes. He wandered aimlessly on.

It was strange how wholly absent were all signs of
human habitation. It was as if he had reached the
end of the world—which was unpeopled. It was true
that thereabouts they might be early folk, but, even
if they had already retired to rest, surely in some
house somewhere a light would be left burning. But
there was not a glimmer; behind, in front, to right
or left, wherever he looked, not one. He had scarcely
realised till then in how bare and bleak a country his
boyhood had, in fact, been spent; how remote from
civilisation, how cut off from the world. It came upon
him, with a sense of shock, how lonely a childhood
his had been. He was returning now to that same
wilderness, a lonely figure, lost in the night.

It was very strange indeed that there should be no
houses.   Used as he had been of late to the life of a
great city, their absence struck him as doubly strange.
No houses!   Where, then, did the people live?   Were
there really none?   Was he the only living creature
environed by the night?   The thing began to weigh
upon him like a presage of evil.   Thereabouts were
the folks all dead?   Was the world peopled only
with ghosts?   Were they walking at his side, peering
at him over the stone walls, scurrying hither and
thither across the country, unheard, unseen?

What was that noise?   It was the sound of waters.
What was that breath upon his cheeks?   Was it not
the whisper of the sea?   What was the sudden, greater
blackness which yawned all at once in front of him,
which made him suddenly stop short, which set him
shivering?

He was standing upon the edge of the cliff—the
Samphire Cliff.   The knowledge of his whereabouts
came upon him with an instant intuition.   Down
below him was the island — that reef of dreadful
rocks.   It was against it the waves were breaking;
that was the noise which he had heard.   Another
step, he would have been over, hurtling through the
air, rebounding from the points of jagged stone, plung-
ing into the whirlpool of the continually contending
waters.   As he started back, he sweated.   His knees
shook.   When he had retreated backwards perhaps a
dozen yards, he subsided on to the ground, trembling.
What an escape he had had!

But it had, at any rate, made him acquainted with
his whereabouts.   There were the home woods, then,
upon his right.   He advanced towards them gingerly,
feeling with outstretched arms for the stone wall
which, he remembered, girdled them.   There it was.
He recognised in it the presence of an old friend.
There used to be a gate farther down the road, per-

haps a hundred yards. This was not the part of the wood which he had been making for. By some accident, which he did not understand, he had struck one end instead of the other. But to have found the woods at all was something. The gate which he was searching for opened into a grassy road which, in the old days, was wont to be used as a drive when a rare visitor was to be taken to enjoy the view from the Samphire Cliff looking towards Godrevy. It was the gate which was farthest from the house. The winding road passed both the upper and the lower lakes, through a part of the grounds which, in bad weather, was very like a swamp.

He found it. The latch emitted a grating sound as he lifted it; the gate shrieked on its rusty hinges as he threw it back. He entered the woodland road. The mystery of the woods was obvious even in the night.

Up there the winds were too persistent to permit of the trees attaining to great proportions. But, as he went in farther, their size increased. If the winds were against them, the ground was on their side. They were sheltered by its rapid fall. Soon Blaise Polhurston knew that he was surrounded by monarchs of the forest. Their branches creaked and moaned; afar off came cries as of souls in pain. Here were living things; he could hear them on every side—the pattering of feet, the scurrying of unseen creatures through the brushwood, the occasional whirring of wings. The wood was far from silent.

There was, too, a presence of another kind. He had not yet wholly recovered from the shock of finding himself on the edge of the abyss. There was still a fluttering at his heart which the place in which he was was not disposed to lessen. It was peopled with memories; the grave in which his childhood was buried. He was like a lost soul, passing through the shades of

what once had been his youth.   Every movement of
the branches—they were never still—was a sigh for
the past, a moan for the present, a groan at the prospect of what was still to come.

The forest was in torment; half affrightedly he
groped through it all.   Behind the trunks of the trees,
over whose roots now and then he stumbled, spectres
glanced out at him—presences which he felt, although
he did not see, shapes which were none the less actual,
but rather the more, because they were incorporeal.
All the way he was haunted by demons.   They grew
thicker as he went.   The place became alive with
them.   He began almost to persuade himself that
he could see them with his bodily eyes; could hear
them with his ears.   They gibbered at him as he
passed.   They touched him.

Was it possible they touched him?   What was that
which brushed against his hair—which shaved his
cheeks—which alighted for an instant on his eye-lids?
Were they fingers?   Of what?   Of whom?   He
stopped suddenly, looking about him with bewildered
glances.   What was that noise?   Was it not laughter?
Who was laughing at him in the shadow of the trees?

This was not to be borne, that there should be foes
on every hand, and he see none of them.   He lit a
match ; he would see, by the light it gave, what might
be seen.   While it flickered, as he was about to raise
it to look round, something—someone struck it out.
He trembled so that the box fell from his grasp.   He
listened.   Again !   Was that not laughter among the
trees?

He could not find the matchbox, although he felt
sure that it had fallen just at his feet.   He could not
feel it anywhere, even by going down on his knees.
Was that it?   If it was, then it was snatched away
—yes, as his fingers were closing on it.   What was
the cause of his feeling that something, someone, was

playing tricks with him? His nerves had been disorganised by his experience on the cliff. The walk, prolonged beyond his expectations, had tired him. Memories had come sweeping over him at an impressionable moment. The surroundings were weird, the scene awesome. They used to tell, in bygone days, of the Polhurston ghost; of the wraith which, at nighttime, in the woods, used to lie in wait for members of the family, when evil loomed over the house.

It was such thoughts as these which bade fair to play the fool with him. He shook himself. He would put them behind him—leave the matchbox where it lay—hasten home. It must be growing late. If he was not quick, his coming would alarm the house—that he did not want.

He strode down the path as rapidly as the darkness would permit. There was the sound of running water. He was nearing the upper lake. Presently, although the blackness shrouded it almost wholly, he knew that he had reached it. He paused by the brink, peering through the shadows. Nothing could be seen. He could hear what he took to be the movements of wild-fowl, the splashing of water over the weir—that was all. He passed on. A closing of the darkness, till it hung about him like a pall, made him conscious that, about the lake, it had been, relatively, light. He was moving again through the thickness of the wood.

All at once, something flashed into sudden radiance; so far as he could judge, at a distance of a hundred yards or more from where he was. It gleamed, like a star, then vanished, leaving the darkness denser than before. What was it? A will-o'-the-wisp? A glow-worm? Hardly, at that season of the year. Besides, it had borne no resemblance to either. He went steadily on. It came again; still, as it seemed, the same distance in front. It was not a light in a house, nothing of the kind. It was like a flash of fire, appearing to him to have the property of showing, in the brief space of its duration, half-a-dozen different hues. He advanced more cautiously, watching for its re-appearance. It did not come.

Instead, there was borne in upon him, on a sudden, an absolute conviction that some unseen thing was close at hand—something which was moving through the brushwood, keeping pace at his side. The instant he became conscious of this assurance, he stopped, and stood, and listened, and looked. He could see nothing, nor hear anything; but he knew that there was something there, and that it had stopped when he stopped. He tried to speak to it, but could not— to move in its direction, but his limbs were chained. With an effort he continued his advance, stumbling blindly on; he knew that the something was moving too. Again there was the sound of running water; he was nearing the lower lake. As he reached it, there was a splash! Something had plunged into the lake. There was a noise as of something struggling in the water—a gasping sound, as of fighting for breath.

"Who are you? Who's there?" cried Mr Polhurston, the tones of his own voice filling him with horror and amazement.

All was still. Only the noise of the weir disturbed the silence.

To have plunged into the blackness of the waters in search of he knew not what would have been an act of madness. If he remembered rightly, quite close to the bank this lower lake was ten or twelve feet deep; while, here and there, were pools of unknown profundity. How did he know what had fallen in—by accident or of set purpose. It might be some creature of the woods. True, it had not sounded as if it were, but—what was it then? He did not stop to think. He rushed onward, resolving, when he reached the house, to tell his tale. In any case, there was nothing he could do, without light, alone.

So soon as he had arrived on the opposite side, and was re-entering the forest's vastness, he again became conscious that still there was something at his side; something, this time. which he could hear— for there was a swishing sound, as if something was being brushed against the trees and bushes as it went. It might have been a trick of his imagination, but, as he tore wildly on, he seemed to detect a dripping sound, as if drops of water were falling from his invisible associate. There was clearly audible, also, a panting, as for breath.

With sudden determination, of which he himself was more than half afraid, swerving from the path, he dashed in the direction from which the sound proceeded. As he advanced it retreated; he could hear it going farther and farther back among the trees. Under such circumstances, to follow it would be absurd. He returned to the path. It returned with him; he heard it come.

He groped about for a stone, or other missile, to hurl at it. Its persistence began to anger him. He could find nothing except broken branches and scraps of decayed wood, which were too light to be of any use. So he assailed it with his tongue.

"Confound you—who are you? What do you want? What do you mean by skulking at my side? Get off, you brute!"

This last he said in view of the possibility that it might be some animal, desirous of human companionship; of which he was very far from being sure. Nothing showed that his words were heard, but when he moved the thing went with him.

"Very good. Wait till we get nearer the house, where there's more light; then we'll see what you are, my friend, if you choose to keep me company so far."

Something laughed. In the laughter was a quality which made Mr Polhurston almost jump out of his skin. He quickened his pace, breaking into a run.

So swiftly, so heedlessly did he go, that before he knew that the thing was there, he dashed against a tree. So violent was the contact, that, partially stunned, he dropped to his knees. It was a second or two before he had regained his senses sufficiently to realise with any clearness what had happened. What aided their return as much as anything else was the discovery that he was surrounded by a sudden illumination. He scrambled to his feet, almost falling again, in his haste, over the tangled root, to find himself confronted by something which made his heart stand still.

It was a girl—a young girl. She stood within three or four feet of where he himself was standing. A sufficiently pretty picture she presented illumined by the sudden, vivid glow. But at sight of her he shrank back, shivering; the blood seemed to freeze in his veins; there was a buzzing sound in his head, as if his brain was bursting.

"Helen!" he gasped. "Helen!"

The girl made some slight movement which caused him to divert his glance; whereupon he perceived that behind her, a little to one side, stood a man, who held her hand in his.

"Shapcott!" muttered Blaise Polhurston. "Helen! —my God!"

It was the woman to whose skirts he had pinned his boyhood's faith; for the love of whom he had wrecked his life. And the man was the friend who had betrayed him; who had hounded the woman into

a suicide's grave, and for the recovery of whose murderer he—Blaise Polhurston—was himself supposed
to be offering a reward of £200.   The vision lasted
but an instant, then it faded.   And in another moment
Mr Polhurston was tearing madly, headlong through
the darkness down the path.

Nor did he check his pace till the sudden widening
of the path showed him that he had entered the main
avenue, which divided the estate, as nearly as possible,
into two clean halves, and about the centre of which
the house itself was placed.   Here some show of reason
did return to him.   He stopped to draw breath and
to wipe the perspiration from his brow.   Then, still
moving quickly, but at a pace which more nearly
approached discretion, he started on the last stage of
his journey.

Almost immediately, however, to stop again.   For,
down a by-path, came a figure bearing a lighted lantern in its hand.   It was the figure of a woman.   She
wore a long cloak, which fell to her feet, the hood
being drawn over her head.   As he advanced she held
up her lantern so that its gleam shone on him; then,
with an exclamation, she moved towards him with
outstretched hand.

"Blaise," she cried.   There was no surprise in her
voice, only pleasure.   "It's you!   I knew you were
coming, so I made haste to meet you."

It was his mother.

Blaise Polhurston had returned home.

CHAPTER XXI

THE GIRL ON THE PATH

IT was odd to find oneself, after all, a person of importance. Mr Polhurston looked about the spacious chamber, in which he had been installed as his own particular apartment, and wondered how long it was since he had been the occupant of a sitting-room. There was a pile of letters on the table at his side. He examined them, feeling, as he did so, as if he were an actor in some tale of topsy turveydom. It was all so very curious. The county had found him out. Here were congratulations, invitations, compliments, pouring in on every side.

His mother entered while he was still engaged with his correspondence. Hers was not the least singular figure in the situation. The discovery of how little she had altered was to him a source of continual amazement. Her hair was white, there were crows' feet about her eyes, wrinkles upon her brow, but in other respects she was the same. The most extraordinary thing was that she treated him as if he were still the same. They might never have been parted, so little notice did she appear to take of the years which had slipped away. That anything might have occurred, during their passage, to make him other than he used to be, such a possibility never seemed to enter into her philosophy. She was the same dominant personality as of yore; she governed the whole house by rule of thumb, and, in the queerest fashion, she included her newly-found son under the ægis of her despotic sway. In the old time, what he had regarded as her tyrannical interference with his righteous liberty had galled him to a raw, now it tickled him immensely.

She had a number of letters and papers in her hand; she wore an air of business. The chair which he offered she declined.

"No, I can't sit down. Blaise, people are waking to a proper appreciation of the situation. Of course, it's only what I expected, still it's satisfactory. Everyone wants me to fix dates for you. I've promised the Verryans you will dine with them on Thursday, and the Tregowans for Friday. There are two or three who write about Saturday, but I thought you might like to have that left open. Then they want you to preside at the market dinners at Camborne on Monday, Redruth on Tuesday, and Truro on Wednesday. I think you might as well. The meeting to choose a new candidate for the division is for to-night. The result, I take it, is a foregone conclusion. I don't know who could be suggested as an alternative for you"—the absurdity of such a supposition made her smile—"so I have given them to understand that you will be prepared to receive the deputation to-morrow. That is a matter about which no time must be lost."

He could have laughed in her face, only his sense of courtesy restrained him. He looked at her to see if she really was in earnest. No one could have been more grave.

"It is very good of you to take all this trouble, but, if you won't mind my saying so, I have returned home, and with your permission I should like to stay."

"It is very nice of you to say so, and I appreciate the compliment. But there is not only myself to be considered."

"No; I also should like to receive some slight consideration."

"You? How you?"

"My dear mother, I have come back to you a little—tired. Thank all these good people for their kindness, and entreat them, for a while, to let me rest."

On her countenance was not only approbation.

"But you must take your proper position in the county."

"My proper position in the county?" He made a little gesture with his hands. "There is time enough for that."

"But about your candidature—that is a matter which admits of no delay; it must be entered on at once."

"As yet I've not been asked to stand."

"But you will be."

"Let's hope I sha'n't."

"Blaise!"

"I'm no politician—not even a little bit of one!"

"What has that to do with it? As Polhurston of Polhurston your proper place is in the House."

"That give me leave to doubt. At least let us wait till I've been asked to stand."

"Blaise! You'll not refuse!"

"Mother, you must allow me some latitude of action. When the proposition is laid down before me in a definite and authoritative form, I will give it my consideration. In the meantime I must beg of you to ask these good people who press on me their invitations to suffer me, for a while, to stand excused. The first essential which I need is rest."

She left him with a shadow on her face. Presently his sister came instead.

This Mrs Hamilton was like, yet unlike, her brother. She had his whimsical eyes and flexible mouth. There was something, too, in her carriage, her gestures, her movements, which reminded one of him. But she was sprightlier than he; more obviously imbued with a knowledge of the world—her own particular world, that is, in which she lived and moved and had her being. As is apt to be the case with women, she was willing to accept, outwardly, conditions against which, inwardly, she chafed. Her philosophy seemed to be that nothing mattered much, so long as friction was avoided, and one obtained to a reasonable extent the things which one desired. That she found her brother a puzzle difficult of solution she made no effort to conceal—especially from him.

She stood before the fire, one side of her skirt a little raised, her left foot on the fender.

" Well, what now ? "

" What now ? A cigar, if I may."

" Have a dozen for me.   What's the shadow on the mother's face ? "

" I hope I have had no hand in its causation."

" But you know you have.   What have you been doing ? "

" I am desirous of doing nothing."

" Does that mean that you want to refuse the invitations she has accepted for you ; that you decline to show yourself to the county ? "

" My dear sister.   I have come back to the mother and to you ; not to the county."

He was leaning back in his arm-chair.   She looked down at him quizzically.

" Do you know, Blaise, that you're a very funny person ?   You won't mind my being frank ? "

" Not in the least.   I have become indurated in the school of candour."

" You have kept yourself away from us all these years, during which mother has been like one wander-

ing in the wilderness, and now that with your return
she has come into her promised land, it is only natural
that she should wish to partake of some of the fruits
thereof."

Mr Polhurston pulled at his cigar for a second or
two in silence. He critically examined the ash.

"My dear sister, I fear that it may not be easy for
the mother and me to arrive at a standpoint of
mutual comprehension, but at least let us understand
each other."

"That's what I want to do."

He had paused, and was again eyeing his cigar.

"Has the mother ever allowed you to do just as
you pleased?"

"Never.   She ruled my childhood, she chose my
husband—you never knew him?"

"Unfortunately, no—alas!"

Her eyes twinkled.

"You need not be so pat with your 'alas!' One
is not necessarily in love with the husband of one's
mother's choice—and then we were married for so
short a time. When he was gone, she bade me come
back home; so I came, and have stayed here ever
since, obedient in all things—that is, with certain
mental reservations."

"I am of a more unruly temperament."

"With you it is different. You never have done
what she wished. She is older than she seems, and
not so strong. You surely can go a little out of your
way to give her pleasure as her life draws to its
close; you would not like her to go to her grave and
leave you with the knowledge that you had never
done anything to occasion her a moment's happiness."

"You think she would not be unwilling to die with
the consciousness that she had never yielded in one
jot or tittle to pleasure me?"

"I see—that is the line which you take up. Well,
P

I'm sorry. I had hoped that your coming home
meant that the past was past."

"You take her part? You think that all the blame
is on my side?"

"Not a bit of it. I think that both sides were to
blame; it's a case for compromise."

"Precisely. I come home—that is my contribution
to the compromise. She should leave me in peace
now that I have come home—that will be hers."

She eyed him shrewdly.

"What is at the back of your mind, Blaise? Why
do you wish to be hidden? Don't you think it's a
natural and excusable instinct on her part to wish
to kill the fatted calf in honour of the returning
prodigal?—to wish to proclaim to all the world that
there is once more a Polhurston at Polhurston, and
one of whom she is not ashamed?"

"I don't know that I do wish to be hidden. I
merely propose not to assume responsibilities for
which I am not prepared; nor to allow myself to be
thrust into an equivocal position. I have been a
person of no importance for so long that I am indis-
posed to reverse the position without reflection—that
is all?"

Mrs Hamilton was looking down into the fire.

"It will break her heart, literally, if you fall foul
of the plans which, all these years, she has been
planning. You must remember that the chief occupation of her life has been to watch, and wait, and pray
for your return. All the time she has been arranging
what she will do to mark the happy hour. Be sure
that your reasons are sufficient before, by pricking all
her schemes, you prove them to be airballs. Dr
Grainger tells me that her heart is affected; which is
one reason why I have been obedient in so many
things."

"Use your influence with her to give me a fort-

night's grace—to suffer me to stay quietly at home
here for fourteen days. Then, at the end of that
time, if all goes well, I will be in her hands the
puppet which she seems to require me to be, apparently with your entire sympathy, and will use my
best endeavour to be the kind of creature she is of
opinion that a Polhurston of Polhurston ought to be."

"Why the fortnight's grace? And what do you
mean by the proviso, if all goes well? What do
you expect will go ill between this and then?"

"My dear sister, in this matter can you not give me
your unquestioning co-operation?"

The brother and the sister met each other eye to
eye, with steady, continuous inquiry. Her glance was
the first to fall, back to the fire.

"I will try to do as you wish, though the task which
you are imposing is not so light as you suppose. And
I hope that all things will go well."

"I hope so too."

"But about the candidature; that requires an immediate decision."

"Then I must refuse."

"Suppose it can be arranged to allow you a fortnight for consideration?"

"If during the fortnight, all things go well, I'll
stand."

"Then, doubly and trebly, I hope that those
mysterious 'all things' will go well. If you do refuse, I believe the disappointment will be greater than
mother can bear; she has set her heart on having a
representative of the family in the House of Commons."

When his sister left him, Mr Polhurston went out to
take the air. As he neared the mill-pond a singular
thing occurred. This mill-pond was associated with
the events which had scarred and maimed his whole
existence. It was a sheet of water of considerable
dimensions; plentifully stocked with coarse fish, the

haunt of wild-fowl; lilies, white and yellow, grew in thick patches here and there, and it was surrounded by a fringe of trees and shrubs. On one side was a low stone wall; by this ran a foot-path; and as in summer the place was not without its charm, it was esteemed a favourite resort by old and young. To approach it from the quarter from which Mr Polhurston was advancing, one passed through a gate. As he reached this gate, Mr Polhurston saw coming towards him along the foot-path a girl.

There was nothing remarkable about her, except that she was pretty, which, where girls are concerned, is so commonplace an accident as hardly to deserve to be called remarkable. She was neatly dressed, with simplicity and taste, which is not too often to be found in persons of the class to which it was plain that she belonged. She bore herself so daintily, with her head and neck well set back upon her shoulders, that she would have been noticeable in any company, if only for her graceful figure and easy carriage.

But still there seemed to be nothing about her appearance to have disconcerted Mr Blaise Polhurston. Yet that he was disturbed, mentally and physically, and that to a serious extent, was plain. He was half through the gate, and had his hand upon it, to hold it open, when she first came within his range of vision. When he saw her he started, as if he was not quite sure what it was he was looking at. He stood still as she approached, staring with all his eyes. As she drew nearer he shrank back as though he feared to come into contact with some dreadful thing; farther and farther back till he was right off the path, and all but tumbling into the ditch. She, on the other hand, appeared to take no notice of him whatever; or, at least, just as little as she possibly could. On her countenance were no signs of recognition. She merely glanced at him, casually, as she came, as one might glance at any stranger. If she observed him at all it was with a slight curling of the lip and head held a little higher, which tokens of resentment, if they were such, were entirely justified by the singularity of his deportment.

He stared at her as if she were some spectral visitant, some being with whose face and form he had been familiar his whole life long—a creature who had haunted him sleeping and waking, who had been the continual tormentor of his dreams. As she went past him, with chin uplifted, and an evident determination not to see the figure cowering on the ditch's edge, he was seized with such a shivering fit that he had to clutch at a sapling to help him to stand. His eyes followed her along the winding path till a turn brought her under cover of the hedge and hid her from his sight. Then, for the first time, he seemed to be able to draw breath.

So complete had been his absorption in the girl's coming and going that he had failed to notice that somebody had followed her down by the mill-pond. It was the vicar of the parish, the Rev. Henry Jardyne, a portly individual, whose religious equipment, one hoped, was of heavier metal than his intellectual.

"Well, Mr Polhurston, and how are you? I trust quite well."

Mr Polhurston snatched at the reverend gentleman's arm.

"Jardyne, who was that who went along the path?"

"You mean the girl?" Mr Jardyne pursed his lips. He shook his head with an air of disapproval. "That's Helen Fowler."

"Who?"

Mr Polhurston, oblivious of his propinquity to the edge of the ditch, loosed hold of the vicar's arm and commenced to slip down the bank with such precipitation, that, if Mr Jardyne had not made a sudden clutch at him, he would have descended into the miry stream.

"My dear sir, take care! Yes, that's Helen Fowler. It's a sad story. Her mother lived in this parish before my time."

"Ah! Before your time?"

"Much before my time, or what took place might not have happened." The vicar's meaning it was not difficult to surmise. "Her mother drowned herself in the mill-pond—this very pond by which we're standing." He protruded his head; he spoke in a whisper. "That girl's her illegitimate daughter."

"She's Helen's child?"

"Yes, Helen Fowler's child. The unfortunate offspring of that wretched woman. And she's not ashamed."

"Why should she be?"

"My dear sir, she's the child of shame."

"Who was her father?"

"As I am informed, a person of the name of Shapcott."

"Shapcott!"

Mr Jardyne did not notice the expression which was on Mr Polhurston's face, being of an unobservant habit, or he might have been surprised.

"I have been given to understand that an individual of that name was loitering in the neighbourhood, and—'Satan finds some mischief still for idle hands to do'—though I am not acquitting the young woman of blame. Of course, what I am telling you I have only been told myself; though, I fear, that there is much more truth in the story than I could wish. Because—to be frank—the girl bids fair to take after her mother."

"What do you mean?"

"She is of an unruly disposition, stubborn, proud to a degree; though goodness only knows what she has to be proud of. She lives all alone in a cottage which was left to her by a male relation of her mother's, himself an undesirable character. Heaven alone can tell what she does for a living. And now I am credibly informed that she has been seen, under most equivocal circumstances, with a person who ought to know better."

Long after the vicar had left him Mr Polhurston remained gazing into the pond, turning things over and over in his mind.

"What's that which Shakspere says about our acts being our shadows—fatal shadows which walk by us still? The seeds which we sow in our youth take seed and grow till we are encompassed by a forest of trees which shroud us in shadow, keep from us the warmth of the sun, and, as it seems, the light of God's mercy. To think that Helen should have a child living and I not know it! Helen's child—and Shapcott's!"

He closed his eyes, and shuddered, as at a picture conjured up by his words. He alone knew what it was.

CHAPTER XXII

A GAME OF BILLIARDS

Miss Dollie Hamilton took what was evidently her own particular cue out of its metal case. She chalked the tip with much precision.

"I suppose that you can play?"

"I used to be able to play a little—once."

"Shall I give you any points?"

There was about this suggestion a touch of arrogance which was unmistakable.

"Thank you. I think we might play for a hundred level just to see how we stand."

He opened with a miss. She replied with another.

"Of course you behaved very badly."

"I'm afraid I did."

"You ought to have told me who you were directly you got into the carriage."

"Before I knew who you were?"

"You ought to have told me directly you did. You ought not to have allowed me to remain in ignorance a moment. It was unmanly."

"Unmanly?"

"Yes. See what I said about you to May Gifford, and what May Gifford said to me, while you said nothing."

"Just so."

"It has placed me in a false position—and her. I don't know what she will say when she knows that you were the man in the corner."

"Perhaps it will be a lesson to you not to discuss third persons in a stranger's presence."

"How absurd!" She brought her cue down on to the floor with a bang. "If I can't talk about my own uncle to my best friend, whom can I talk about? How am I to know that my uncle likes to play the part of eavesdropper?"

"How could I play the part of eavesdropper when I was there all the time in plain sight?"

"But not in plain knowledge."

"I think you're severe. The circumstances were peculiar."

"Most peculiar. But I don't see how that makes the position any better for you. All the same I'll forgive you, since you are my uncle—that is, if you'd like me to."

"I should—very much."

"Then you're forgiven. That's a fluke." He had screwed a cannon in rather ingenious fashion and did something very like the same thing three or four times over. "Of course, if you are going to play like that, you ought to have let me know. I ought to have had points. I'm not a professional player."

He sighed.

"How many points shall I give you?"

"None, thank you, in the middle of a game. If the balls will only run decently for me perhaps I may make a break. I happen to-day to be in a forgiving mood."

"I hope that's not peculiar to to-day."

"Oh, but it is; I'm not forgiving as a rule, I assure you. I haven't forgiven you before, have I?"

"No, I'm afraid not. But better late than never."

"But to-day I've had good news; Jack's got a brief!"

"I'm very glad to hear it. Who's Jack?"

"Who's Jack? Mr John Armitage is the gentleman to whom I am engaged."

"Engaged! You don't mean to say that you're engaged?"

"And why not? Why do you look at me like that?"

"You—you seem so young."

"Young! I seem young! My good uncle, I'm positively ancient."

"Are you? Then what am I?"

"You! Compared to me, you are positively juvenile. At least, you seem so to me. A woman is always older than a man."

"Is she? I see. Now I understand."

"Now that he's made his first step up the ladder I trust that people will be more reasonable, especially mamma. There ought to be a formal announcement. Because, of course, if he wins this case, he'll get shoals of others."

"Not a doubt of it."

"As Jack himself says, it's the sort of case in which a man has an excellent opportunity of bringing himself before the public eye. He's lucky to have got it—though, no doubt, he can't be long among even the dullest people without their discovering that he's something very much out of the common."

"So I should imagine. In what sort of case has he been briefed?"   ·

"It's a case of murder."

"Oh." With much deliberation, Mr Polhurston essayed a long jenny, and scored. "A case of murder."

"It's very odd, at least it seems odd, and a little embarrassing. But there's always something."

"In what way?"

"So I thought I'd be the first to tell you."

"To tell me what? That Jack's got a brief?"

"It's not only that—that is, not exactly. I hope you won't be annoyed."

"Annoyed? What at! At Jack's getting a brief? I hope he'll get a thousand."

"Well, you see"—she was re-chalking her cue with much assiduity—"what makes it seem rather awkward is that it's in the case of Mr Shapcott."

"Of—" He was about to make a stroke, but, at the mention of his name, he paused and turned. "Of whom?"

"Of Mr Shapcott—your Mr Shapcott."

"Oh." He made his stroke, and scored. "That does seem rather odd."

"I understand that you've offered a reward for the detection of the criminal; and, of course, you'd like to see the man who did it hung."

"Well—to continue?"

He scored again.

"Jack's for the defence."

"Is that Mr Armitage your Mr Armitage?"

"You knew he was for the defence?"

"I understood that a Mr Armitage was to be counsel for the prisoner, and I was told that he was a very clever man."

"Who told you? Do tell me who told you?"

Her eyes sparkled. She moved towards him with eager steps.

"Oh—someone. I'm not sure that I quite remember who it was, but I know that I was told."

"You mustn't be angry if he wins."

"If he wins?"

"If he gets the prisoner off. He has to do his best, for his own sake, and—and for mine."

"And for the prisoner's?"

"Yes, of course—and for the prisoner's. He may be innocent."

"He may be."

"You won't be angry if he gets him off, even—even if he's guilty?   You see Jack's so very clever that he may get him off in any case.   Think what that would mean to him, and—and to me!"

"My dear Dolly, can you keep a secret?"

"Can I keep a secret!   As if I couldn't!   Have I breathed a word to any living soul about how you behaved in the railway carriage?   And I'm going to bind May over to inviolate secrecy."

"That's very good of you; but this is—rather different."

"That doesn't matter; I can keep all kinds.   I've kept dozens—that is, I've kept all I've ever had to keep."

Taking his glass out of his eye, he carefully wiped it with his pocket-handkerchief, looking at her with a smile which suggested peculiar enjoyment of the situation.

"I want there to be a private understanding between you and me, of which you're not to breathe a word to any living soul."

"What is the understanding to be about?"

"If Mr Armitage procures the prisoner's acquittal, I'll give you—a set of diamonds?"

"You'll give me a set of diamonds?"

"Something like a set."

"If Jack wins?"

"That's it—if Jack wins."

"But—I thought you wanted the prisoner to be hung."

"Why should you think it?"

"You've been offering a reward for his discovery."

"A reward has been offered for the discovery of the actual murderer.   The prisoner may be innocent."

"How do you know he's innocent?"

He turned to the table.

"I believe it's my stroke.   I think I'll try to

double the red into the middle pocket." He doubled it. "I didn't say I did know that he was innocent. I said he might be."

"Then why are you so anxious for his acquittal? He may be guilty."

"The prisoner's name is Robert Foster. He is a young man, a very young man; like yourself, a positive ancient. I have had inquiries made about him; he seems to be an honest, hard-working, self-conducted youngster; a strong impression of his innocence has been left upon my mind."

"Then why do you prosecute?"

"I don't; it is the Crown."

"But you can go into the witness-box and say what you know."

"I can."

"Then why don't you?"

He fenced with her question, although she did not notice it.

"In England one cannot go into the witness-box merely to express one's opinion. It's your stroke."

She played, and missed, then turned eagerly to him again.

"But I don't understand. From what I knew, and from what Jack says, I had so taken it for granted that your heart was set on a conviction that now I know it isn't, my ideas seem to be all disarranged. Tell me—why are you so anxious for an acquittal?"

"Aren't we all desirous that innocence should be made plain?"

"That's all very well, but when it comes to a set of diamonds!"

"You think that my anxiety for abstract justice goes too far? Very good. Shall I withdraw my offer?"

"Not at all—I think it's a very proper present to make me in any case."

"That's kind of you."

"Suppose he's not acquitted?"

"Then, according to the letter of the bond, the offer's off; you'll get no diamonds."

"Oh!" Her face was a vivid note of exclamation. "I think that's very hard on me. I can't compel them to let him go."

"Perhaps Mr Armitage may be able to bring about the desired consummation."

"Jack's not omnipotent. You don't seem to remember that this is the very first brief he's ever had."

"Are you suggesting that it would have been wiser to have elected a counsel of more experience?"

"The idea!"

"If you are of opinion that the matter would have been safer in more experienced hands, it is perhaps not yet too late to alter the arrangements which have been made."

"How horrid you are! If Jack can't win, no one can—you may be certain of that. But even Jack can't be sure of winning, can he? Are you ever going to leave off scoring? There, that's game. I think you ought to have told me before we began that you were a splendid player, then there would have been something like a proper handicap; I've hardly begun."

"There's still time for your proper handicap. Shall we have another game?"

"I don't care. But tell me, never mind my diamonds, what shall you do if he isn't acquitted?"

"He will be."

"I'm not so sure. Jack himself says the case against him's very strong; he says that one girl's evidence alone is almost enough to hang him. Her name's Hills—Pollie Hills."

Mr Polhurston was manipulating the scoring board with his back towards Dollie.

"How many points shall I give you?"

"Sixty, at least."

"That's a lot."

"I ought to have seventy; you play better than Jack, and he gives me eighty."

"Does he ever win?"

"Well, sometimes — when he's disagreeable, he does."

"And I suppose when you're agreeable he loses."

"That's one way of putting it, no doubt. Do you know I'm beginning to suspect that you're sarcastic? that you hardly ever intend what you say? that nearly all your speeches have a double meaning? That's a kind of character I particularly dislike."

"I'm sorry that I should come within the definition of the kind of character towards which you have such a natural aversion."

"You're not a bit sorry, you're laughing at me now. But I don't care, I'll tell you something else, and you can laugh at that. I believe you're a mystery—that your whole life's a mystery; like one of those tales which come out in penny numbers, and keep on coming out for years and years because the mystery wants such a frightful lot of unravelling."

"Perhaps a similar good fortune may attend on me. I believe that the lives of noteworthy criminals are occasionally published."

"Are you a noteworthy criminal?"

He shrugged his shoulders.

"I beat you the last game."

"Yes, and if you beat me this, you'll be a criminal of the very deepest dye. No excuse will be accepted. I'll take seventy instead of sixty; you're to start. You're not to give a miss—you're to play at the red, and you're to leave both balls over the pockets—you haven't told me what you'll do if the man's found guilty."

"How do you mean, what shall I do?"

"Shall you let him hang?"

"What steps do you mean me to take to prevent it—cut the rope?"

"Do you mean to say that you will let him hang when you know he's innocent?"

"I did not say I knew that he was innocent."

"But you do know—I'm sure of it."

To him there seemed to be something significant in her tone. He cut the red into the middle pocket; then he looked at her.

"You are sure of it?"

"Don't you know that he is innocent?"

She appealed to him with a little imperative gesture. He chalked his cue, smiling as he did so.

"My dear Dollie, you appear to attribute to me a variety of qualities. I'm a man of mystery, and now, it seems, I'm a man of knowledge too. To be able to positively affirm that this man was innocent, I should require to know who was guilty."

"Don't you?"

"Don't I what?"

"Don't you know who killed Mr Shapcott?"

He still smiled, as he glanced at the tip of his cue to see if the process of chalking was complete.

"You flatter me by supposing that if I did I should have kept the knowledge to myself. It would have made me accessory after the fact."

She impatiently brushed a truant lock back from her ear.

"All I can say is, I don't understand it in the least. Directly I heard what Jack was going to do, I made up my mind that there was going to be a pretty complication; for I felt sure that you wouldn't relish the idea of his getting the prisoner off, and now it seems that it's all the other way. What I feel is, if you don't know that the man is innocent, why are you so anxious that he should be acquitted? And if you do know, why don't you say so? It almost looks as if you didn't want to pay that £200 reward you offered."

"You think so? You mustn't always trust to looks. It's your turn. Now you've got the balls, you ought to run right out."

"I don't care to play any more, thank you; I've had enough. I'm no match for you; you can win whenever you please—I believe if you gave me ninety-nine out of a hundred—and I don't care to be made a laughing-stock. If you don't mind I think I'll go and write to Jack. I must write to him to-day, and this may be my only chance. I'll tell him what you have told me."

"Do!"

"Though I don't in the least understand what it is you have told me. But I can only do my best to make things plainer to him than they are to me."

"Beyond doubt in your efforts to achieve lucidity you will be successful. You will bring to the task the sufficient qualification of a calm and well-balanced mind."

She looked at him; she replaced her cue in the rack; and she left the room in something very nearly approaching to a pet.

When he was left alone Mr Polhurston laughed to himself, as if in the enjoyment of some private and particular joke; and he tried his hand at some nursery cannons.

CHAPTER XXIII

A VOICE FROM THE OTHER SIDE

In front of the house was a muddy pond, partially filled with water, whose surface was coated with green slime. On one side of this was a hedge. Between the pond and the 'hedge was a narrow, ill-kept foot-path which led to a broken clapper gate. Within this gate was what had once been a garden of fair size, but which was now little less than a piece of waste land, so overgrown was it with weeds and thistles. About the centre of this piece of ground was the house itself, a one-storeyed erection, built of stone, with a slate roof.

As Mr Polhurston stood at the gate he felt that he had never seen a more unpromising abode. Everywhere were the marks of ruin and decay. From the appearance of things, it was years since a spade had been put into the ground. A small outbuilding, which stood on one side of the house, and which had, possibly, once been a cowshed, was falling to pieces. Of what had once been a pig-sty, nothing was left but rotting timbers.

The house itself stood badly in need of repairs. The one chimney pot was a shattered wreck. Slates were missing from the roof, green moss covered most of those which were left; while on one side flourished a patch of vigorous nettles. It was difficult to believe that anyone could be living in so neglected a dwelling-place, and when one reflected on its aloofness from other human habitations, that the nearest house was almost a mile away, and had to be reached through narrow, winding lanes, which, at this season of the year, mud rendered almost impenetrable—and remembered that it was the abode of a young and pretty girl, who was its sole occupant, one could but hope, for her own sake, that she was fond of solitude, that her nerves were in good order, and that she had ways and means of her own for preventing the time hanging too heavily on her hands.

The place seemed so utterly deserted, and so wholly void of signs of life, that, as he picked his way between the thistles which hemmed in the narrow path, Mr Polhurston concluded that the girl herself must be away. He was, therefore, the more surprised when, almost as soon as his knuckles touched the panel, which stood badly in need of repainting, the door opened, and the girl stood confronting him on the threshold.

The girl's startling resemblance to the one who was gone, the suddenness with which she had opened the door, these things, joined to a consciousness of the delicate nature of his errand, and to his retrospective and even somewhat morbid mood, tended to disconcert him. He stood hesitating, as if doubtful what to say. It was she who addressed him.

"Well? What do you want?"

"I wish to speak to you. May I come in?"

She drew back, so as to permit him to enter, then closed the door. His awkwardness had not deserted him even after his entrance. He still continued tongue-tied, staring at her as if she were some spectre, instead of a creature of flesh and blood. It was again she who broke the silence.

"I knew you would come."

Even her voice was another's. He stared about him like a man in a dream—dimly realising that whatever the house might appear without, within it was a model of neatness and order. The scanty furniture was polished till it shone again. The stone floor was spotlessly clean. The kitchen utensils were examples of unremitting care. Nowhere was there a suspicion of dust. Some half-dozen books were on a shelf at one side. The only ornament which the room contained was the cabinet portrait of a young man, which hung against the wall.

Seeing the preoccupation of his mood she repeated what she had said, as if she supposed it had been unheeded.

"I knew you would come!"

Her voice caused him to glance at her again. As he did so, it was as if a flood of memory swept over him. He seemed to be looking, in very deed and very truth, at his boyhood's love. The resemblance was little short of marvellous. It was not only an affair of form and feature, but of expression, pose, a dozen little tricks of speech and gesture which were as fresh in his recollection as on the day on which he had seen them last. A wild, unreasonable desire sprang up within him to take her in his arms and strain her to his breast, but it was a desire which was not to be encouraged, for—for some reason which, as yet, was strange to him—in her bearing there was hardness, scorn, almost rage, which was far from suggesting a caress.

"You knew that I should come?"

"Yes. I knew it."

"Then you know who I am?"

"I should think I do know. I ought to. You are Mr Blaise Polhurston."

"I am. I trust that you also know that I was a friend of your mother's."

In his voice there was a note of tenderness which was very far from finding an echo in hers.

"Her friend? Her enemy, you mean."

"Her enemy? God forbid! That I never was."

"Your mother and you hounded her to death between you."

"You are under some strange delusion. I would have given my life to have saved hers. Indeed, I have given her the best part of it as it is."

"You were not only my mother's enemy, you are also mine!"

"Yours? I am your enemy? You dream!"

Her manner of addressing him was so entirely different to anything which he had expected, that he was taken more and more aback.

"Do you think that I know nothing of what goes on in the world? You credit me with too much ignorance. Do you think that I don't know that my father—my father—" she repeated the word with added emphasis, "has been murdered, and that you have all his money, and that I, his child, have none? You hounded my mother to her death—for all I know you have had a hand in the killing of my father—you take the inheritance which should be mine, and then, when I say you are my enemy, you exclaim that I dream. What, then, in your judgment, goes to the making of an enemy?"

He sank on to a chair which was at his back. Her words were so unexpected, the charges with which she assailed him so unlooked for, that, in the first flush of his amazement, he could but stare—a fact on which she placed her own interpretation.

"Does your conscience prick you that you stare at me as if I were a ghost? I had supposed, from what I have heard of you, that you had got beyond the stage in which you would be likely to find a conscience a troublesome possession."

"You are mistaken. My conscience still troubles me more than sufficiently at times, though not for that which you suggest. There it's clear. Until the last few hours I did not know that a child of your mother's was living, and still less, that Howard Shapcott was your father."

"Oh, yes, he was my father. I'm a child of shame, doubly and trebly. My mother committed suicide, my father was murdered; what can I do to complete the tragedy? All my life fingers have been pointed at me. I've been told that I'm a girl without a name, a thing apart, a creature marked with an ineffaceable brand."

"You are sure you don't exaggerate?"

"Exaggerate!" She laughed—not a pleasant laugh to hear issuing from a young girl's lips. "Ask the parish if I exaggerate. Ask the first person on whom you come if I'm not Helen Fowler's child; if my mother didn't drown herself because she couldn't face the shame of having borne me; if I don't call myself by my mother's name, only because I've no right to that of any other; if I haven't been the cause of all the trouble at the great house; and if there's any better prospect for me than to do as my mother did, or worse. My mother's folk threw those things in my face so long as any of them were living, and now that they are gone, what is there for me to do but to stay here, all alone, and think of them, morning, noon and night?"

"You could hardly have chosen a more unhealthy occupation. No wonder your mind is filled with such imaginings. A young girl ought not to live alone in such a place as this."

She laughed again.

"Where else should I live? Under what other roof should I seek shelter? This place is my own. And as for being alone in it, I'd rather. I never go

out in the daytime where people are. They look at
me, and thrust each other in the side, and say, 'There
she goes! That's Helen Fowler's shame!' When I
want to go out I wait till after dark, so that I can
have for company the creatures of the night. They
don't point at me and tell each other who I am."

There was an air of wildness about the fashion in
which she said this, which was not without a pathos
of its own. It went so ill with her beauty and her
youth.

"But how do you live? Have you means?"

"No, I've no means—you've all my father had."
Again there came that jarring laugh. "It's little
enough I want in the way of food. And as for
clothes they last me a long time, since I see so little
company." As she spoke Mr Polhurston could not
help but observe that the print dress which she wore
seemed new, and how tastefully it was fashioned,
and how well it became her.

"The little money I want I earn by sewing. I go
far afield for work, where folks don't know me;
they trust me with it, and let me bring it home to
do it here."

There was a pause before Mr Polhurston spoke
again. He chose his words with such care as he
could, for the feeling was strong upon him that the
situation was more delicate even than he had supposed.

"I cannot but think that the story of my acquaintance with your mother, whom I loved, esteemed, and
honoured, and would, if I could, have made my
wife—"

"You would have made her your wife?"

"If I could—indeed! I cannot but think that this
story has been told you by some prejudiced person,
to the serious disadvantage of my share in it. I
must beg you, for the moment, to take my word for

it, when I say that I would have done all I could for your mother, just as now, I will do all I can for her child. As you observe, I have your father's money. As I have already told you, until quite recently I did not know that Howard Shapcott was your father. Why he made me his heir I cannot tell you; I had never any reason to suppose that he intended doing anything of the kind. But now that I am acquainted with the truth, you may rest assured that I will see that justice is done."

"What do you mean by saying that you will see that justice is done?"

"You pay me a poor compliment by asking such a question. Surely my meaning must be sufficiently obvious. I had no claim on Howard Shapcott—you are his child."

She shrank away.

"He has never been to me a father; it is only from what they've told me that I know he was my father. He never treated me as his child; I never saw him. I don't want his money now that he is dead."

"You, at least, have a better right to it than I, as you yourself have pointed out."

"They were only my wild words; I didn't mean it."

She drew farther back from him, as if fearful that he would proffer her the money there and then.

"Do not be afraid. We shall not quarrel, you and I. Everything shall be done in due and proper order. My trust is that you may learn to know me better, and like me none the less with increased knowledge. What we have to do at present, you and I, is to arrange some plan by which you can remove with the least delay from this lonely cottage, and take up your residence in some place better suited to your age and situation. My opinion is, that the farther away you go—within reason, that is—the better."

"I don't want to leave this house."

"You must. On that point there can be no two opinions. It is not right that a young girl should live, like a solitary hermit, all alone by herself, at the other end of nowhere."

It was plain that the decision with which he spoke, blended, as it chanced to be, with something which was at the same time both humorous and tender, impressed the girl even against her will. A chord seemed to be touched which had hitherto remained quiescent in her bosom. She exclaimed, with apparent sudden impulse,—

"I will go and fetch a letter which I have to give you from my mother."

Before he could reply, she had vanished into the adjoining room. Her words, wholly unforeseen, and to him mysterious, had thrilled him. He sat in a state of suspensive doubt, expectant of he knew not what. Even yet he had not been able to shake himself wholly free of the feeling that he had returned, at least in part, to the days of long ago; that he had taken up, as it were, a thread which had long been dropped, and in so doing had been restored to the presence of his old love in a new form. The incident had for him almost a flavour of the supernatural. He had lived so long with and for a ghost that now it almost seemed as if, all at once, the ghost had been clothed with flesh, and filled with the breath of life. He found it impossible to shake himself free of a preposterous half-suspicion that this girl was the girl whom he had loved—the very one—with whom time and circumstance had played some trick for the delusion of his senses. So that when she spoke of giving him a letter from her mother, as if she were close at hand to write it, it seemed to be the one touch which was needed to complete the confusion of his brain.

Presently the girl returned, bearing an envelope in her hand. She spoke with a catching in her breath, as if in a tremor of excitement; as she spoke, in spite of the agitation of his mind, he was constrained to perceive, as in a sort of mist, how pretty she looked, and how like her who was gone.

"They found this after—mother—went. When I grew up, they gave it me. I've kept it ever since. It's for you."

She delivered to him the envelope; he accepted it with shaking hand. He held it for a moment, silently, a haze seeming to gather before his eyes. When he could, he looked at it. It was yellow with age. On it was something written; the ink had grown dim. The writing itself was of that careful yet uncertain sort which betokens the unpractised scribe.

"Let the child have this, to give to Blaise Polhurston when she sees him, as she will some day."

That was what was written on the envelope. As he read, he held his breath. Then he opened the envelope, and took out what was within. On a sheet of common paper, written with a bad pen and worse ink, there was this letter:—

"Dear Blaise,—You've been very good to me, though I've been very bad to you. I've always loved you best, though I've been so wicked. I couldn't help it; he made me. I believe your mother set him on. But I know that's no excuse, so I'm going to make an end. I know you'll forgive me for having been so wicked, because I know you would forgive me anything. So I want you to help the child if ever she needs help—which she's sure to, because he'll do nothing, nor give a pennyworth more than he is forced. I want you to do it, dear Blaise, for my sake, although she's his. I wish she was yours, then I shouldn't be doing this, because, God knows, I love you best, and always have. So I am, dear Blaise, in spite of all, your loving   Helen."

That was the letter, a poor screed, after writing which it appeared that the woman straightway went and drowned herself. So that it seemed to be, in more than a merely figurative sense, a voice from the other side.

This man, with his queer composite nature, had made of the woman an idol, for whose sake he had esteemed the world well lost. He read the letter through, hearing in it her voice speaking to him, seeing in it her very soul, the very heart of her. After that one reading, the mere words were indelibly printed on his brain, to keep always in plain sight; and, with them, the message it conveyed of love, weakness, trust—that combination of dissonant qualities which had made her what she was. So that, for the present, the one reading was enough.

He gave the letter to the girl.

"There," he said, "is my authority to help you."

He laid his head on the bare boards of the deal table, encircling it with his arms, so as to conceal his face. He was still, while the girl, standing at his side, in her turn read her mother's letter, with, on her pretty countenance, wonderment writ large.

CHAPTER XXIV

THE DEPUTATION

His sister met him on his return to the house. She had evidently been waiting his arrival, because, directly he appeared round the bend, so as to bring him within line of the door, she came hastening down the avenue. A golf-cap was thrown over her head. She was in what, for her, was a state of excitement.

"Blaise, the deputation's come."

"The deputation?"

He glanced at her out of the corners of his eyes, as she fell in by his side.

"Yes, the deputation—about your standing for the division. If seems that Mr Glenthorne's very ill—it's perhaps only a question of hours, and there's a feeling among the people that something ought to be settled at once—at any rate, provisionally, especially as the other side have got their candidate, who's already in the field. So they've come over to press you to become the member—which is what it amounts to. Why do you smile like that?"

"Mayn't I smile?"

"Oh, you may smile, so long as you do it in the proper way. But that's not a proper way. What's the matter with you? Is anything wrong?"

"Is it likely when all's for the best in this best of all possible worlds?"

"Pray, don't be cheaply cynical. If you must go in for that kind of thing, at least try to raise your style a little above the common ruck. It sounds smarter, and, in that sort of talk, sound does mean so much. There's Colonel Liddell inside; he says he met you in town."

"On a memorable occasion. I remember that it was he who first spoke to me of Philip Glenthorne."

"He's very keen that you should stand, but, of course, that's only natural; they all are."

"Most natural."

It was her turn to peep at him askance; but, if she suspected a second meaning in his words, she allowed no hint of her suspicion to escape her.

"Then there's George Treevennack, and John Eva of Redruth, and Sir Charles Rayner, and the Vicar—five in all."

"It will give me great pleasure to meet these gentlemen."

They had reached the house. As they entered the hall, Mrs Hamilton turned and faced him.

"I hope you're not going to be silly, Blaise, or worse? Mother's in there."

"Mother is in there?—I will bear that well in mind."

As he entered the room they greeted him with outstretched hands. It was plain from their demeanour that they were already of opinion that the object of their errand would be gained. Possibly Mrs Polhurston had had something to do with their state of assurance. They clustered round him as if greeting one whose victory was assured; that he would enter the lists as a contestant they plainly had no doubt. Mrs Polhurston, seated in a big chair in the background, beamed on the proceedings as if they were being conducted under her personal superintendence.

"So you have come at last, Blaise. We have been waiting for you."

"I was engaged elsewhere."

There was a significant dryness in his tones which escaped her notice, and theirs.

"I have asked these gentlemen to give us the pleasure of their society at lunch. As it is now so late, perhaps we had better have luncheon first."

Colonel Liddell dissented—with what he meant to be a display of humour.

"No! Ten thousand pardons, Mrs Polhurston, but to that suggestion I am afraid that I must say positively, no. This is a case of business first and pleasure afterwards. Because, suppose—I only say, suppose, since, as men of business it's our bounden duty to consider all possibilities, however remote they may appear to be, and, I trust, in this instance, are—I say, suppose he were to refuse to accede to our request, with what faces could we sit down with him to table? With what hearts? What appetites?"

Mr Polhurston interposed.

"I hope, gentlemen, that you will allow no considerations of the kind to prevent my mother having the pleasure of your society. By so doing you will render my position more difficult than it already is. Because, while fully sensible of the compliment you would pay me, I regret that it is impossible I should do as you desire."

Blank silence followed his words, broken only by an exclamation from his sister.

"Blaise!"

Presently Colonel Liddell found his voice.

"My dear Polhurston, don't you think you're rather premature in announcing your decision before you've afforded us an opportunity of explaining what it is we've come about?"

"I understand that you have come to ask me

to offer myself as a candidate for the parliamentary division."

"That is so; and if you will allow us to place the matter before you in our own way—"

"Excuse me, Liddell, but I would spare you so much trouble. Nothing you could say would alter the position. I repeat that it is impossible for me to stand. So impossible that I must decline to enter on a discussion."

"Polhurston!"

The Colonel stared at him as if he could not believe his eyes and ears. Mr Treevennack spoke.

"May I ask, Mr Polhurston, when you arrived at so definite and uncompromising a resolution?"

"When Colonel Liddell broached the subject to me in town, I told him that it would be impossible for me to stand."

The Colonel fidgeted.

"I supposed that that was merely a form of speech—that you would require a formal invitation."

"You were not entitled to such a supposition. My words were clear and final."

The Colonel flung out his arms.

"But why should it be impossible? What are your reasons?"

"At this moment I am unable to give you my reasons, or any reasons. I think it probable that before very long you will learn them for yourselves. I can only say they are sufficient."

Sir Charles Rayner put in his word.

"But you are placing us in a difficult position—in a position, in fact, which is likely to turn out almost unendurable. You are leaving us without, so far as I know, the shadow of a shade of an alternative candidate, and Glenthorne may be dead to-morrow. We have been allowed to take it for granted that you would come forward; indeed, up to the moment of your coming into the room, your mother was giving us to understand—"

"Be so good as not to bring my mother's name into the discussion."

"Upon my word!" spluttered Sir Charles.

He was of a generous figure and a choleric disposition, and Blaise Polhurston's tone was not exactly genial.

Mrs Polhurston entered the breach. She had risen from her chair and had drawn herself erect, like some high priestess whose authority was momentarily questioned by a fractious insubordinate.

"But, my dear Blaise, you must stand. I have set my heart upon your standing."

"I can only regret, mother, that you should have set your heart upon what is unattainable."

"Unattainable! Such language is foolish. You have only to come forward to be elected; nothing could be more simple. Leave the matter to me, Blaise; I will see that all is as it should be."

"Do you, then, propose to come forward as a candidate?"

"Blaise!"

"Because I can only repeat that I do not. And, since these gentlemen hear me, I can only suppose that they will hardly be so foolish as to announce a candidate to the constituency who refuses to stand, and who, even if elected, would refuse to take his seat."

"Blaise! What exaggerated language you use! You must not listen to him, gentlemen; you will find that, after all, he is amenable to reason. Come, let us go into the dining-room. We will talk to him while we're at luncheon; you'll see if I'm not right."

"I am afraid, mother, that you must allow me to stand excused. I have some very pressing letters

which I must write. I have the pleasure, gentlemen, of wishing you good-day. I can only express
my regret that I find myself unworthy to fill the
high position you propose for me, though, at the
same time, I'm very sure that you'll have no difficulty
in finding an, in every way, more desirable candidate
than I could be."

He moved towards the door. His mother's im-
perious voice checked him.

" Blaise! Where are you going ? "

" To write some letters, which must be written."

" Stay. I wish you to lunch with us."

" I am sorry, mother, that I cannot."

Without giving her an opportunity to interpose
another word he left the room. Without he encountered Miss Dolly Hamilton, who waved a pink
paper about her head.

" I've had a telegram from Jack."

" I hope it brings good news."

" Well, some of it's good and some of it's bad.
The Grand Jury has found a true bill against Robert
Foster."

" I'm sorry to hear it."

" I thought you would be. That's the bad news ;
the trial's to be on Monday."

" Monday ? That gives me three clear days. It
ought to be sufficient."

" It's sufficient for Jack to come down and have a
peep at me. He's coming to-night. That's the good
news ; and isn't it good ? "

" To-night ? Mr Armitage is coming here to-night ?
Then I will stay."

" Of course you'll stay. Where were you going ? "

" To write some letters."

With a nod and a smile Mr Polhurston walked
away, and left her staring after him with the telegram in her hand.

R

CHAPTER XXV

THE TWO WOMEN AND THE MAN

BLAISE POLHURSTON looked up from his writing at the sound of the opening door. It was his sister who had entered.

"Blaise, mother wishes to speak to you."

"Mother?" Although he looked at her, it seemed to be with unseeing eyes, as though he was turning something over in his mind. "Do you think that any good purpose will be served?"

"Cannot your mother speak to you when she wishes, even though no good purpose my be served, that is, from your own special point of view? It seems evident that it is only to your own point of view that you intend to give consideration."

He paid no heed to her closing words. He rose from his seat with a little gesture of acquiescence.

"I suppose that it may as well be in that way as in any other."

"What do you mean?" Without waiting for him to reply she followed with another question. "Blaise, have you any natural feeling?"

He smiled at her; what she was beginning to feel was his exasperating smile.

"My dear sister, don't let us discuss abstractions. Where is my mother?"

She looked at him as if she found him beyond her comprehension ; which, indeed, she did.

258

"You are aware that you have inflicted on her a bitter disappointment; stultified her, indeed, in the eyes of those before whom she is little accustomed to be abased. Why you have done it you alone can say; it is a mystery to her, and to me. You must excuse her, therefore, if she uses bitter language. Try to remember that she's your mother."

He planted himself immediately in front of her, looking at her with a half-amused scrutiny, straight in the face.

"Since I entered this house I seem to have returned to school. You two women appear to consider yourselves entitled to exercise over my movements a directive control. You favour me with little lectures on what you hold to be the rules and deportment of polite society, and when I venture to act on my own initiative, you resent it. Give me leave to tell you, with all possible politeness, that what I do, I do; and that for what I do, I hold myself accountable to myself alone. Now, may I ask you to conduct me to my mother?"

Without a word she turned and went out. He followed her. She paused for a moment outside the door of Mrs Polhurston's room, as if, in spite of what he had said, she found it impossible to avoid giving him a final warning.

"Spare your mother as much as you can; she has been used to play the schoolmistress so long."

Mrs Polhurston sat in a straight-backed chair, beside a table which was littered with papers. This was the apartment in which she transacted the business of the estate. It contained abundant evidences that she was not merely an ornamental mistress. Her son, as he looked at her, recognised, not for the first time, that she was the true, great lady of a former school. It was with a feeling which was not unmingled with amusement that he realised that he stood half in awe of her, as if he had been guilty of some conduct for which he feared to receive at least a scolding.   Nor was this impression lessened by the air of silent reproof with which she regarded him, as if what he had done was too heinous to be lightly spoken of.   It was he who broke the silence, when he began to find that it was becoming too trying.

"I am told that you wish to speak to me."

She continued silent for still some seconds longer. When she did speak, her manner was cold, severe, formal.

"You know I wish to speak to you.   I want an explanation of your extraordinary behaviour."

"Since I am unconscious of extraordinary behaviour I might feign ignorance of your meaning; but that I won't do.   You mean that you want to know why I decline to stand as a candidate for the division.   I believe that you will find one of my reasons sufficient. To-morrow I am leaving Polhurston."

"You are leaving Polhurston?   What do you mean?"

"I should have gone to-day; only I understand that Mr Armitage is expected here to-night.   As I particularly wish to see him, with your permission, I will stay to do so.   But to-morrow I go."

"You go?   I presume you mean for a day or two?"

"I mean for ever."

"For ever!   Blaise!"

His sister struck him with impatient fingers on the arm.

"Blaise!   How can you be so wicked?"

"When I tell you what I intend to do I think that you will yourselves condemn me to perpetual banishment.   I propose, if God is willing, to make a home for Helen Fowler's child; in which, I trust, she will look upon me as her guardian, or, if I am so fortunate as to gain her confidence and affection, her father."

The two women looked at each other as if in doubt whether it was they or he whose senses had fled.

"Helen Fowler's child? You mean that — that bastard?"

Nothing could have exceeded the bitterness of Mrs Polhurston's tone. Her son suffered it to pass unnoticed.

"Precisely. Howard Shapcott's daughter. You will remember that at present I have his money, which should, of right, be hers."

"Hers! Blaise, you're mad! She was never a daughter to Howard Shapcott. He never set eyes on her—or wished to. It was never even certainly proved that she was his child."

"Mother!"

"In any case, he did all that was required of him. He gave her mother's relations a hundred pounds, more than the strict justice of the case demanded.

"You and I are not likely to see this matter with a single eye, therefore, mother, do not let our last hours together be embittered by the sort of words which are likely to be spoken if we dwell on it. Let me go in peace and not in anger, so that my banishment may not be so perpetual as I have feared, and now and then I may be able again to come and see you. I have some letters which I must finish, but in half an hour it will give me great pleasure to place myself at your disposal."

He made as if to leave the room; she stopped him.

"Blaise! You don't mean it?"

"Don't you know, mother, that I mean what I say? Have you forgotten?"

"But—but it's sheer midsummer madness! The girl's a wastrel!"

"Not yet. With God's permission she shall never be."

"She is, I tell you! She's misbehaved herself with young Angel of Treswithian."

"Mother, I implore you not to speak to me like that. Let us spare each other. Let me go now and come back in half an hour, when you've had time to think things over."

"What do you imagine I've been doing all these years but think, think, think? The girl came between us all that time ago, and spoilt your life and mine. And now is her nameless brat to come again —now that I am old, and you no longer young? Blaise, I entreat you not to leave me! You're my son—my only son—all that I have! All day long, right through the years, I've watched for you, through the weary nights I've listened for your footsteps; now the day has dawned, and you've come home; stay with me, Blaise, stay with me— till I go home for ever. It won't be long—won't you stay with me a little while?"

Mrs Polhurston's reserve was at an end. The heat within had dissolved the frozen exterior. The passion which she had kept hidden away had escaped from its hiding place and burst to the front. Her reversion to nature caused her to present a pitiful spectacle of emotional old age. Her son seemed to be at a loss for words with which to answer her.

"I will come and see you whenever you wish."

"You will come! But why should you go? Why should you go?"

"You will hardly agree to my bringing Helen Fowler's daughter here?"

"Here! Here! That girl! Blaise, have pity on me. What is that girl to you? I am your mother! She is the living evidence of her mother's falsehood —of her wanton nature."

"Can you not spare me these speeches? Have you not learnt sufficient wisdom? They did no good in the time gone by. Do you suppose that now they will do more? I want to keep back the words which are tingling on my tongue—don't goad me. Help me!"

Mrs Hamilton interposed.

"Come, mother, calm yourself. I am sure that if you and Blaise talk over matters quietly together you will arrive at a satisfactory understanding; you will find that he is conscious that he has other ties which he ought to consider besides those of which he talks. Now, Blaise, what is it that you propose?—and think before you speak."

"You don't allow me much time for thought, but here's my proposition. I must go away to-morrow—at any rate for a few days. During my absence I will find a home for the girl, and then, if all goes well, I will return, at least for a while, so that we may arrive at a better understanding."

"Why do you say, if all goes well? Do you anticipate that anything is likely to prevent your return?" He was silent, looking down at the eyeglass which he was polishing. "Ever since you have come back, Blaise, I have felt that there is something at the back of your mind; something which weighs on you; which hampers your freedom of movement. Come, Blaise, tell us what it is—tell mother and me. We are your nearest, if not your dearest. You must yourself perceive that it is not fair of you to allow us to face, unprepared and unprotected, a danger of whose very nature we are ignorant."

"My dear sister, you don't sufficiently allow for the singularity, from your point of view, of my life. Events have come crowding on me. Only the other day I was a pauper, with scarcely a shoe to my foot—"

" A pauper ?   Blaise ! "

The interruption came from Mrs Polhurston.

" I must apologise, mother, for touching on so unpleasant a subject, but the thing is true. The change in my circumstances has come about with pantomimic suddenness. Even yet I have not clearly realised why and how it's happened. You are probably aware that on Monday a lad is to stand his trial for the murder of Howard Shapcott. I need not tell you that that is a matter in which I take the keenest interest."

" Why should you ?   It is not one with which you are actually concerned."

" On the verdict my movements will materially depend."

" How can that be ?   What verdict do you want ? "

" A verdict of acquittal."

" Of acquittal !   But if he's guilty ? "

" I have no evidence which I am in a position to deduce, but I have reason to know that he is not guilty. You would not have me allow an innocent man to hang ? "

" You know !   Blaise !   How can you know ? "

He shrugged his shoulders.

" There are some matters, my dear sister, into which it is not wise to inquire too curiously."

As Mrs Hamilton watched him she seemed to perceive something on his face which caused the fashion of her own countenance to change. She stood by Mrs Polhurston's chair. Stretching out her arm she took her mother's hand in hers, and held it fast. The two women stared at the man, and all three were still.

CHAPTER XXVI

JOHN ARMITAGE

MISS DOLLY HAMILTON'S Jack turned out to be one of the old young men who are not the least characteristic features of the day. He had very fair hair, and not much of it, smooth cheeks, a quiet manner, and a pair of greyish eyes, which struck one as being the soul of discretion.

Not that his precocity was of the sort which suggested that the wine of life ran low; or that, as regards quality, it was thin. On the contrary, he was an athlete — big, both as regards height and across the chest, with long arms. And though he conveyed the impression that everything which he did was done leisurely, when he moved one was surprised to find how fast he went. Before he had been in his company five minutes, Mr Polhurston came to the conclusion that Miss Dolly had not made so bad a choice as she might have done. This was a man.

The dinner hung fire. Mrs Polhurston was cold and silent; she seemed, indeed, to be far from well. Mrs Hamilton was fidgety—a very unusual thing for her to be; one might have said that self-possession was her strongest point. Dolly was nervous; something which was in the air had the effect of stopping the flow of her naturally irresistible high spirits.

No one, probably, was sorry when the meal was at
an end.

It was still the custom at Polhurston for the ladies
to leave the gentlemen at table. When the two
men were left alone, Mr Polhurston filled his glass,
and, raising it to his lips, looked over the brim at his
companion.

" I understand, Mr Armitage, that you are pro-
fessionally engaged in a case in which I am interested ? "

Mr Polhurston had been conscious throughout
the meal not only that the scrutiny with which
he had been observing the young man had been,
though unostentatiously, returned in kind, but he
also had a shrewd suspicion that, despite the dis-
parity in years, this young man was probably to
the full as keen a judge of character as he himself
could claim to be. As he put the question the
greyish eyes, turning towards him, seemed to fasten
themselves upon his face.

" You mean the Queen against Foster ? "

" I do. I suppose, indeed I know, that very much
more attractive mettle awaits you elsewhere than
anything which I can offer ; yet I shall esteem it a
very great favour if you will give me answers to
some questions before you—go to it. I may tell
you that it is to put these questions to you that I
have postponed my departure from this house until
to-morrow."

Mr Armitage inclined his head.

" What I am about to say to you will be said in
the strictest confidence, and I must ask you to make
no allusion to anything which may pass between us
to a living creature outside this room."

Again John Armitage inclined his head. Mr
Polhurston would have preferred him to speak, yet
he appreciated his discretion.

"I don't know if you are aware that I am providing the funds for the prisoner's—that is, for your client's—defence?"

"I was not aware of it until this moment." He paused, continuing to look at Blaise Polhurston, on whose face, or in whose eyes, for it was a case of glance for glance, he seemed all at once to see something which inspired him with a sudden decision to be as frank as it was plain the other desired. "I may add that I have myself been a little curious as to the quarter from which they did come. It is obvious that Foster has not the necessary means; and, in fact, he himself seems to be as much in the dark as I have been."

"He does not know who pays?"

"Not only does he not know who pays, but I am inclined to the opinion that he would rather no one did pay. I believe that, if he had his own way, he would dispense with the services of both solicitor and advocate. He would prefer to be undefended."

"Is that so?" Mr Polhurston appeared to be examining the contents of his glass—his companion continued to examine him. "What is the impression on your mind as to his innocence or guilt?"

"I don't know that I could answer that question, even if I were sure it is a fair one—which I'm not. But I don't mind telling you that if matters go as at present they bid fair to go, the jury will bring in a verdict of guilty."

"That is your opinion?"

"That is my conviction."

"What makes you so positive?"

"The man's own behaviour. He seems determined to hang himself."

"To hang himself?"

There was an interval of silence before Mr Armitage

spoke again, during which the two men again met each other eye to eye.

"As you probably know, this is my first case of the least importance. Therefore you may say that I am not qualified to judge; but I am persuaded that though I have hundreds, nay, thousands, more, I shall not have one in some respects more curious. You appear to be interested in the man, or you would not provide the funds for his defence. I take it that you know something of his story. It is possible that you may be able to throw light on certain points which are beyond my comprehension—light will have to come from somewhere if he is to be saved from the gallows."

"What are the points to which you allude?"

"Do you know what answer he gives to every question which Maguire—that's his solicitor, you know—and I put to him? 'Pollie shall hang me.'"

"What?"

"'Pollie shall hang me.'"

"What does he mean?"

"So far as I can understand, this. One of the chief witnesses against him is a girl named Hills—Pollie Hills. She alleges that he gave her a ring and a bracelet which were admittedly the property of the murdered man, and which were supposed to have been stolen on the night of the murder. It was she who betrayed him to the police. He says that just before she did so he told her that he loved her, and asked her to be his wife; and was endeavouring, as he supposed, to save her from the police when she caused him to be arrested. His line of reasoning seems to be that, as he told her of his love, and her reply was his betrayal, life is not worth living; since it is plain she hates him, and that, therefore, she may as well be allowed to carry her hatred to a logical conclusion, and hang him out of hand. So it happens that, whenever he is asked for an explanation or a statement his one formula is, 'Pollie shall hang me;' meaning that, since the girl wants to hang him, why, he intends to let her—so there's an end."

"Is the girl's evidence material?"

"Most material. If what she says is true, hardly anything could make things look worse for Foster."

"Do you think it is true?"

Mr Polhurston was still examining his wine. John Armitage seemed to consider before he answered, his grey eyes always on the other's face.

"I hardly know what to think. If the man would only open his mouth I might have something to go upon, but, in the face of his persistent silence, things are at a deadlock. I admit too, that, as matters stand, I see no reason why the girl should commit a serious and apparently motiveless perjury. That someone gave her the stolen property seems certain. If not Foster, then who? There is nothing to show. And remember he himself does not deny it. What do you know of Robert Foster?"

"Nothing which will be likely to be of assistance to you. I have been given to understand that he is an honest, hard-working, decent, well-behaved young fellow, who earns his living by hawking vegetables, not the sort of person to commit the crime with which he is charged. Under the circumstances, therefore, I felt it to be my duty to see that adequate arrangements were made for his defence. And I feel sure that he is in good hands; Dolly tells me so."

"Dolly!" John Armitage's face brightened, as if the name sounded pleasantly to his ear. There was an obvious pause. Then he put another question. "What do you know about the girl?"

Looking up from his wineglass, Blaise Polhurston again met him eye to eye.

"Again nothing which will be likely to be of assistance to you."

"But what do you know?"

"I have given you my answer."

The lawyer's glance returned to his glass. He
sipped its contents. In some subtle fashion one
gathered that he had learned something from the
other's answer which was not upon the surface, and
that Blaise Polhurston knew he had. Presently the
latter spoke again.

"And is this girl's the only material evidence for
the prosecution?"

"Unfortunately, no. Or he would not stand in
such imminent peril of being hanged. His connection with the revolver is a hard nut to crack. The
revolver which was found lying by the murdered
man, and with which the crime had evidently been
committed, was his. That's damning."

"What proof is there that it was his?"

"There are his initials on the weapon. Then the
man has come forward from whom he purchased it.
Others assert that they saw it in his possession so
recently as the day on which the deed was done."

"Who are they who assert that?"

"One man is a hawker. He says that Foster
showed him the weapon while they were waiting in
Covent Garden Market on the morning of the same
day."

"He is sure it was the same?"

"He swears it. Another man swears that he saw
him hawking in the streets that same afternoon, that
he spoke to him, and that he took the revolver out
of his pocket and showed it to him. Foster appears
to have been proud of being the owner of such a
weapon, and in the pride of possession to have
exhibited it to every person who could be induced
to look at it."

"Armitage, as a lawyer and a man of the world, what is your estimate of the value of human testimony!"

"Your question goes a long way. My answer is that it depends. If a statement which is sworn to in the box cannot be shaken, that is legal evidence, and, so far as that particular case is concerned, good enough for me."

"Is that so? How odd. And what does Foster say to what these men assert?"

"Nothing. He only repeats the parrot cry, 'Pollie shall hang me.'"

"So that that young woman appears to be at the bottom of all the mischief."

"She's the beginning, the middle, and the end. I am told that she's sorry she spoke; that the police have had trouble with her; that's she's made at least one attempt to abscond. It seems to have been a case of first thoughts are the worst—of doing first and considering afterwards. She'll be sorrier before she's done. It's quite possible that there'll be a scene in court when she's brought face to face with the man whom she betrayed."

There was a dryness about the lawyer's tone which induced his companion to smile.

"Most possible. Who knows? What mere man shall predict, with any degree of certainty, the working of a woman's mind? And is that the whole case against him—the girl's evidence and the revolver?"

"The Crown has pieced the case together very neatly. The chinks have been filled up in a way which we shall find disconcerting. For instance, they are going to show that about the date of the murder he was in possession of considerable funds. If they were not the proceeds of his crime, it will be for us to prove where he got them from. In face of his pig-headed dumbness, that's more than we are likely to be able to do. Of course, there's always the possibility that he's holding his tongue because he esteems that to be better policy, knowing that to speak would be to give himself away. But, somehow, that's an explanation the correctness of which I doubt—and Maguire doubts it too. Do you think he's guilty?"

"I think as you do."

"That his silence is actuated by Quixotic motives, and that he could go some distance towards clearing himself if he chose?"

"You express my meaning exactly." There was another interval of silence. Mr Polhurston refilled his glass, while John Armitage kept his eyes upon his face. The younger man seemed to find his senior an interesting study. "Then, am I to understand that you think that the Crown will get a verdict?"

"As matters stand, I am sure of it. Of course I will do my best, but as things are going I am facing fearful odds."

"I am sorry to hear it—very sorry, very sorry indeed." He looked up. I had hoped that Maguire and you would have managed to procure the acquittal of Robert Foster."

"I only wish we could. No man likes to lose a case, especially when it's his first. But, personally, I feel as if I were fighting with my hands tied behind me."

"Have courage, Armitage. Crippled men have fought to a win before to-day. I hope you may have better success than you anticipate, with all my heart."

The door opened. Miss Dolly Hamilton appeared.

"May I come in?" Mr Polhurston rose to greet her. "Are you two going to talk all night? Do you know that it's frightfully dull for me? How long am I to be without a soul to speak to?"

"No longer. It is my fault that you have remained in that undesirable state so long. I was just wishing Armitage success on Monday."

"Oh!" The girl looked from one man to the other then spoke to Armitage. "Do you know that Uncle Blaise has promised me a set of diamonds if you win?"

"If I win!"

"Permit me to modify the terms of my undertaking. You shall have your diamonds, Dolly, win or lose."

With a nod and a smile he left the room. The girl turned to Mr Armitage.

"What do you think of him?"

The young man looked down at the tablecloth, on which he was tracing figures with an empty wineglass. He was silent for a second or two, then seemed to speak from the fulness of his heart.

"I think he is a man whom the world has not used very well, and will probably use worse. I am no astrologer, nor have I essayed his horoscope, yet I will hazard a guess that he would be found to have been born under an unlucky star."

Perhaps she was in a fanciful mood, but it seemed to Dolly that there was an ominous ring about her lover's words. She shuddered. She never forgot the picture which Blaise Polhurston presented as, with a characteristic inclination of his head, he smiled at her as he left the room.

In the morning he was gone. He had quitted the house before the family was up.

s

CHAPTER XXVII

FOR HIS LIFE

THE judge had taken his seat, the jury had been sworn. There ensued an interval, during which the clerk rose and had a whispered discussion with the judge. He had a blue paper in his hand, which he opened so that the judge could hold half of it. Its contents were evidently the theme of their remarks. Presently the clerk resumed his seat. The judge sat up, settling his robe on his right shoulder, and looking about him through his glasses, as if he was not precisely pleased to see that there were so many people present.

"Bring in Robert Foster!"

Robert Foster was brought in, with a warder in front and another behind. All faces were turned to him. The people at the back craned over each other's shoulders to get a glimpse at his features. The judge gave a twitch to his spectacles, peering at him with an expression of severity.

The prisoner did not seem conscious of the interest he roused, or else he was indifferent. He had neglected his appearance to the full extent permitted by the prison authorities. His hair was unkempt and ragged, his cheeks unshaven, his eyes were bloodshot, his expression lowering. He looked dirty, as if he had not washed for weeks. His clothes were soiled, creased, ill-fitting, as if they had been put on anyhow. He wore no collar or necktie; his greasy grey flannel shirt was open at the neck. He was unlike the brisk, spruce Bob who used to be. One briefless barrister whispered to his fellow,—

"Looks as if he were predestined to fill a niche in the Chamber of Horrors!"

His colleague nodded and made a sketch of the figure in the dock for reproduction in an illustrated paper.

The charge was recited, the prisoner paying not the slightest heed. He looked at the counsels' tables, and then round the court, as if in search of someone; then at the jury, as if he wondered who they were, and was not impressed by their appearance, anyhow. Then he looked at the judge, the sight of whom appeared, for some occult reason, to amuse him—a fact which it was plain that that high dignitary not only perceived but resented.

"Do you plead guilty or not guilty?"

The clerk had to repeat the inquiry before the prisoner could be brought to understand that it was being addressed to him. When he did he hesitated, as if considering what answer he should give. Counsel turned to look at him; John Armitage, in particular, endeavoured to catch his eye—in vain.

"Oh, go on, make an end of it—hang me right away! Guilty!"

John Armitage sprang to his feet. Leaning over the bar, he whispered to his client, who listened to what he had to say with sufficiently unwilling ears. Yet he yielded to what his counsel had to urge, although ungracefully enough. Mr Armitage endeavoured to explain.

"I am for the defence. My client wishes to alter his plea!"

"Do you plead guilty or not guilty?"

"Not guilty—it makes no odds to me !"

The words were scarcely respectful, the manner was defiant. There was a slight commotion in the court; people murmured to each other, staring at the prisoner, who still seemed indifferent to their glances. His solicitor whispered to his counsel,—

"That handicaps you pretty severely. If you get him out of it after that the days of the miracles will have returned."

John Armitage was still; he kept his eyes fixed on his brief. The case for the Crown was opened by that eminent silk, Sir Haselton Jardine. Sir Haselton did not play to the gallery; he played to win. What he had to say was audible to those whose ears it was intended to reach; he was indifferent as to whether he was heard by the occupants of the court in general, which did not tend to make his speech exciting. He told a plain tale, plainly, leaning, it would seem, neither to the one side nor to the other; indeed he expressly told the jury that so far as he was concerned they were at full liberty to take any view of the facts they chose.

"Jardine's got a dead sure thing," murmured Gregory Maguire to John Armitage, "or he wouldn't talk like that. He's made up his mind that our man's going to have no show, and I'm afraid that you won't be able to get him one."

John Armitage was silent.

On a certain morning John Howard Shapcott was found dead in his chambers. It would be shown that he had been killed by a shot wound, which was unlikely to have been self-inflicted. A revolver, one of whose chambers had been discharged, was found near the corpse. It would be proved that this revolver was the property of the prisoner, and had been seen in his possession on the day of the murder. There were signs which suggested that a struggle had taken place and certain property had been stolen, which the jury would hear had been given by the prisoner to a girl named Hills a very short time after the crime had been committed.

It would also be shown that about that period he was in possession of considerable sums of money. Medical and other testimony would be placed before the jury, which would point to the hours between nine o'clock and midnight as those during which the deed was done. Witnesses would be put into the box who would swear to having seen the prisoner in close proximity to the dead man's chambers during the hours named.

It would be for the defence to explain these facts. It was possible that they were capable of an interpretation other than that which they seemed to suggest, in which case it was for the defence to tell them what that interpretation was. If they could succeed in showing that they were consistent with the prisoner's innocence, counsel for the Crown would be glad.

That, in effect, was what Sir Haselton had to observe, and said it as simply, nay, as baldly, as he easily could have done. Yet in spite of his studied appearance of open-mindedness, the impression left upon his hearers was that, if he abstained from using strong language, or laying the colours on thickly, it was merely because it was not necessary to do anything of the kind; since nothing could be more obvious than the prisoner's guilt, as would be clearly shown to the entire satisfaction of every reasonable creature; and how strongly he felt that everyone who heard him was a reasonable being, he need not say.

Sir Haselton called his witnesses. George Armour was the first. He was a hall porter at Embankment Chambers—a tall, grizzle-headed man, with the sign manual of the old soldier written large all over him. He explained that, puzzled at not receiving the usual reply to his knocking at Mr Shapcott's door, he had gone on to the balcony, and, looking through the window, had seen that gentleman lying on the floor. Forcing his way in, he had found that the tenant of the rooms was dead.

Then came police evidence, describing the position in which the body was lying, and the condition of the room. There were signs of a scuffle; a chair and small table had been overturned, and broken ornaments and other articles were lying on the floor. A revolver was lying near the body—the weapon which was produced. On the stock were the initials 'R. F.' One of the barrels had been discharged. An empty envelope was also found, which, according to an endorsement which was on it, had contained a bracelet. The envelope in question was handed to the judge, and from him to the jury. Three drawers in a writing table were open, whose contents had been apparently hastily overhauled.

Cross-examined by John Armitage, Inspector Gardner admitted that no attempt seemed to have been made to rob the corpse, except that a ring was missing which he was known to have been in the habit of wearing, which ring, interpolated Sir Haselton, was in the hands of the prosecution, and would be produced in due course. His gold watch and chain had not been touched; a letter case, containing bank notes, was intact in his coat pocket; he also had upon him gold and silver coins. And there were jewelled links in his shirt cuffs, besides many objects of value about the room—jewelled trinkets, gold and silver plate, numerous trifles of much intrinsic worth, which were extremely portable, and might easily have been converted into cash. If the motive of the crime was robbery, what suggestion had Inspector Gardner to offer why these valuables should have been left behind? Inspector Gardner had none.

"They could hardly have escaped the murderer's notice?"

"I should say not unless he was in too much haste."

"You mean haste to get away after the murder had been committed?"

"Yes."

"But what reason was there why he should not have noticed them before?"

"I cannot tell you. I should say that he must have noticed them. I cannot say why he left them behind, unless it was because he was suddenly disturbed, or alarmed."

"Were there any signs of such disturbance or alarm?"

"I saw none."

"With the exception of the overturned table and chair, the apartment was in its normal condition?"

"I should say so."

"Now, I put it to you, Inspector Gardner, whether this table and chair were not quite close to where the body was lying; and whether it was not possible that they had been overturned by Mr Shapcott?"

"Quite possible."

The judge wanted to know how far from the body they were.

The inspector explained that the table was about three feet. It was small, round, of ebony, very light, and might have rolled after it had fallen. Portions of one of the broken ornaments were found beneath the body itself. The chair was about a foot away, at the back.

"How do you say," continued Mr Armitage, "that the murderer made his entrance and his exit? We have heard nothing of that."

"We have been unable to discover. The door was locked on the inside, so that he could scarcely have got out that way.   There were two French windows, both of which were found closed."

"So it amounts to this, that the man was found dead in the room, with the door locked inside and the windows fastened also inside, so as to have rendered it impossible for anyone to have got either in or out?"

"Certainly not impossible."

"What is the possibility you suggest?"

"He might have gone through one of the windows. They shut with a clasp; he might have pulled it to after him."

"And then, what could he have done? The window is at a considerable height from the ground?"

"That is so.   He could certainly not have jumped. Nor, so far as it has been ascertained, could he have made his escape through either of the other rooms which opened on to the balcony. Some of them were occupied at the time. The windows of those which were not were fastened, the doors locked, and the keys in possession of their tenants. None of them had been burglariously entered. In short, it is a puzzle to explain how the murderer could have effected his escape."

"Don't you think that, under these circumstances, the facts point to suicide rather than to murder?"

"That is not for me to say. I can only tell you what the facts are."

"Precisely.   And the fact is, that it is impossible to even frame a plausible theory to explain how anyone could have got into, and still less out of, Mr Shapcott's room?"

The inspector grudgingly admitted that he supposed that that was what the thing amounted to. He was subjected to a little more badgering—by John Armitage, by the judge, and then, again, by Sir Haselton Jardine.

The next witness was Samuel Miles. He was a pawnbroker's assistant. He remembered selling the revolver which was handed to him; he remembered it very well. He sold it to the prisoner. On that point he had no doubt whatever. He picked him out at once from among a number of other men. Mr Armitage failed to shake this witness's testimony.

William Barlow came next. He was a general dealer. He recollected the date which counsel mentioned—the day of the murder. That morning he was at Covent Garden trying to pick up a bit of likely stock. Prisoner was there on the same errand. They talked together while they waited. Prisoner showed him a revolver which he had purchased; it was the one which counsel handed to witness. He was sure it was the same weapon. He examined it closely at the time, and prisoner particularly pointed out his initials, which he had scratched on it. John Armitage found this witness another hard nut. He stuck to his date, declaring that he remembered it because that morning he purchased a barrel of apples, more than half of whose contents he discovered, when he came to open it, were rotten.

Thomas Bradnock followed. He also was a general dealer. He, too, carried in his memory the day in question. About noon he was with a barrow of fish in Rose Gardens, Hampstead, when he met the prisoner, who was hawking vegetables. They stopped and chatted. During their chat prisoner showed him a revolver which he had purchased. It was the one produced. Thomas Bradnock also John Armitage failed to shake on the question of the date. He held to it that the conversation alluded to took place on the day of the murder.

After Thomas Bradnock came Peter Wood. Mr Wood swore that two nights after the murder he was in the Bull and Ram, a public-house just near Smithfield Market, kept by Mr Scroggins. Prisoner was there, and some other blokes. They had a little discussion about guessing the weight of a bullock. Prisoner wanted to bet five pounds that he could guess as near to it as witness. Witness said that prisoner hadn't got no five pounds, whereupon prisoner pulled out a handful of gold. Witness should think that there was forty or fifty pounds. He was astonished. He had known prisoner pretty near all his life, and had never before known him to have so much as a pound—not all at one time. He asked him where he'd got it from. Prisoner said he'd had a bit of luck; but he wouldn't say what it was, and when they pressed him, prisoner got his temper up and went away.

William House, market porter, was present on the same occasion. He corroborated Peter Wood. Mr Armitage was unable to shake the evidence of these two witnesses on any essential point.

"Call Pollie Hills," directed the junior counsel for the Crown, the solid Mr Pickering.

"Pollie Hills!" cried the usher.

She was not in court. The word was passed without to bring her in.

CHAPTER XXVIII

THE VOICE AT THE BACK OF THE COURT

POLLIE HILLS was accompanied into court by a constable. He kept close to her shoulder, escorting her through the people to the witness-box. There was a look upon her face which suggested that, for very little, she would have torn and rent him; he wore an air of watchfulness and caution as if of opinion that, at any moment, she might mark her estimation of his character by some action of the kind.

Up to that point the prisoner had displayed but faint interest in the proceedings. Now and then he had smiled, as if amused at something which was being said, but for the most part he had looked as if, no matter how it might be with others, the proceedings bored him, and he wished that they would make haste and get them through.

On the appearance of the girl his demeanour changed. He had wakened up at the mention of her name. He had turned in the dock, and, watching her entrance into court, had followed her progress towards the witness-box. Now he kept his eyes fixed upon her, as if he could not remove them if he would; from the working of his countenance it would seem as if he were being torn, mentally and physically, by a torrent of conflicting emotion. She, on the other hand, studiously averted her glance from the dock, never once peeping at the man whose life was

283

already hanging on a thread; yet one felt that she was conscious of his presence in every fibre of her being, and that, quite probably, his gaze was playing havoc with her nervous system, as if it had been a red-hot knife.

Sir Haselton Jardine took her in hand himself, and found her not an easy witness to manage.

Her name was Pollie Hills. Never mind what she was. That was no business of his. Yes, she had seen the ring which was handed to her before, likewise the bracelet. Of course she had. Hadn't they been hers?

" Where did you get them ? "

" Found 'em."

" What do you say ? "

" Say I found 'em."

Sir Haselton regarded the witness in silence for a moment. His tone was drier than usual.

" Please to remember you are on your oath. Since, I take it, you don't wish to commit perjury, I would recommend you to reconsider that reply. I ask you again—where did you get that ring and bracelet ? "

" Blowed if I'll tell you."

The girl was gripping the ledge in front of her with both hands. She was staring into space, as if something which she saw there had induced her to screw her courage to the sticking point.

The judge interposed.

" If you don't take care, witness, you will get yourself into serious trouble. If your purpose is to serve the prisoner you will probably do him more harm than good by attempted evasion. Repeat your question, Sir Haselton."

" Where did you get this ring and bracelet ? "

" They were give me."

" By whom ? "

"Him."

She nodded the back of her head towards the dock.

"Do you mean the prisoner?"

"Course I do."

While the brief contest was taking place between witness and counsel, the prisoner clung to the rail of the dock as if he were hanging on the witness's words. Now that the required answer had been obtained, he lounged back, his hands in his pockets.

"My Gawd!" he muttered, in a voice which was audible all over the court. "My Gawd!"

She turned on him like a wild cat.

"I can't help it. You've brought it on yourself! Why couldn't you keep your nose out of what was no concern of yours?"

"Order! Order!" cried the usher. And the judge pointed out that it was contrary to all rule and custom for prisoner and witness to bandy words. But neither was in time to prevent the prisoner's reply.

"All right, Pollie, you can hang me if you like."

The examination was continued like some game of cross questions and crooked answers. The girl tried her utmost to evade the inquiries which were pressed upon her, replying to them in as vague, loose, and misleading a manner as on the spur of the moment she was able to. More than once she came into conflict with both judge and counsel. Yet her struggles were worse than vain. The impression she left upon the minds of her auditors was that she had been endeavouring, at all and any cost, to screen a criminal from the well-merited consequences of his notorious guilt, and that the criminal was Robert Foster.

John Armitage, however, seemed to be of a different opinion. He rose to cross-examine.

"I believe that you have known the prisoner some time?"

"A good time."

" Has he ever been your lover ? "

" Never."

" Has he ever been on terms of familiarity with you ? "

" Not he—nothing like."

"That you swear ? "

" I do."

" He has been merely an ordinary acquaintance ? "

" That's it."

" Then how came he to give you this ring and bracelet ? "

" I don't know ; you had better ask him."

"I am asking you—and I intend to be answered. But first of all, you will describe to the court, in detail, the exact circumstances under which these things were given you."

She had a bad time with Mr Armitage. He involved her in a mass of contradictions. It was some time before she perceived his drift. Before she had clearly realised that it was his intention, if possible, to convict her out of her own mouth of having lied, she had already made more than one bad slip. When she understood her danger she made an effort to pull herself together ; then she gave the story a more consistent form. But, on the whole, John Armitage succeeded with her much better than he had done with either of her predecessors—so well, indeed, as actually to incur his client's resentment.

When, having been summoned to the prisoner's side, he was expecting to receive some possibly important suggestion, which might be of material assistance to him in his cross-examination, he was astonished to have, instead, addressed to him this gruff and even threatening inquiry,—

" Why can't you leave the girl alone ? "

John Armitage looked his client straight in the face, evidently taking him aback with his rejoinder.

"That girl's lying. You know she's lying, and I know she's lying; and I intend that the judge and jury shall know it too, before I've done with her. If I can help it, I don't mean that you shall be indulged in your taste for suicide."

He did not succeed in laying the girl's perjury perfectly clear, not so clear as he would like to have done. But he did manage to create an atmosphere of distrust about everything which she had said. And when, finally, he suffered her to leave the witness-box, she left a strong feeling behind her that complete reliance could not be placed on any statement she had made. It was felt that there was some story about her connection with those incriminating articles of jewellery which had not been made plain, and that she had devoted all her energies to prevent its being made plain.

"If I could only have got the whole truth out of her," observed Armitage, as he resumed his seat, to Mr Maguire, the solicitor who had instructed him, "the case would have been at an end. Unless I'm mistaken, she holds the key of the puzzle in her hands. I believe that she, and she alone, knows who killed Howard Shapcott."

"In that case she's gone out of the witness-box still the sole repository of that knowledge, which means she's beaten you."

"She has beaten me; I admit it. But one reason for that is that my client's with her and against me; and a pretty strong reason it is; confound the fellow!"

Henry Baynes, solicitor, of Lincoln's Inn Fields, informed the court that the dead man was his client, and the acquaintance of a lifetime. He recognised the ring and bracelet produced as having been the property of the deceased. He was emphatically of opinion that Howard Shapcott was

an unlikely man to commit suicide. He knew no
reason why he should have been guilty of such an
act.

Gregory Lysaght, of Tipton Court, Devonshire,
and Brook Street, London, who stated that he had
probably been the dead man's most intimate friend,
corroborated Mr Baynes. He also recognised the
ring and the bracelet. He was quite certain that
Howard Shapcott would have allowed neither of
them to quit his custody while it was in his power
to prevent it. As for suicide he scouted the idea.
They had lunched together only that morning, when
they had arranged to go to Cairo together after the
New Year. Mr Armitage made no serious attempt to
cross-examine either of these witnesses.

Two men swore that they were on the Embankment, close to Embankment Chambers, on the night
of the murder, and that between ten and eleven
o'clock the prisoner passed them. He was carrying
something in his hand, and walking very fast. One
of these men, who was an acquaintance of his, called
out to him as he went by; but, instead of stopping,
he quickened his pace, nearly breaking into a run.
One of them remarked to the other that he moved
"as if the coppers was after him."

Then came the medical testimony, which went to
show that while it was not actually impossible for
this to have been a case of suicide, it was, to say
the least, extremely improbable. The doctors were
strongly of opinion that murder had been done.

When it came to Mr Armitage's turn to produce
his witnesses for the defence, it was found that he
had none. Instead, he dilated with much skill on
the weakness of the case for the Crown. He pointed
out that all the evidence was circumstantial, that
much of it was flimsy, that some was untrustworthy.
For instance, it was impossible to give much credit

to the statements of the girl Hills. It was not for him to sort out the false from the true, and to do so would be beyond his capacity; but, after her behaviour in the box, and her self-contradictions, it would be doing a serious injustice to his client to attach importance to anything she had said.

Then, again, the manner in which other of the witnesses, after a considerable lapse of time, had attached dates to casual conversations and trivial occurrences was nothing short of extraordinary. They remembered, with precision which was truly astonishing, the most insignificant details which had an apparent bearing on the case in hand, while he had shown, out of their own mouths, that they were far from carrying this habit of exactitude into the important personal concerns of their own daily lives.

In conclusion, he desired the jury to observe that, even supposing all that had been said by the witnesses for the Crown was true — and, for himself, very much of it was marked incredible—it did not necessarily point to his client's guilt. Very much the contrary.

It was admitted, even by the police, and the Crown's own witnesses, that Robert Foster's general character was good; that he was honest, sober, industrious. There was nothing whatever to show why he should have so suddenly and wantonly broken loose from the whole tenour of his life. Was it suggested that, being filled with an instant instinct for murder, he had run amuck in search of a victim? The suggestion was absurd. It was not asserted that he had any sort of acquaintance with the dead man. Why, then, should he have set out to murder him?

Counsel for the defence had no hesitation in expressing a strong opinion that this crime, if crime it was—which had not been conclusively established—

T

had been committed by someone who had an intimate knowledge of the peculiarities of Embankment
Chambers. Everything pointed to that conclusion.
There was the question of entrance and egress. How
had anyone got either in or out? It was an admitted puzzle. Until it was shown that under the
circumstances anyone could have got either in or out,
he asked the jury to accept the alternative of suicide.
But, in any case, it was in the highest degree improbable that such a person as his client could have
been capable of such a feat of legerdemain.

Murder must be proved up the hilt. Before a man
could be condemned to the gallows his guilt must be
established beyond the possibility of doubt. The
mere suggestion that there could be no doubt about
his client's guilt was puerile. There were all sorts of
doubts. And while a shred of one of them remained,
it was their simple bounden duty to give this man
the benefit. Therefore he confidently left his client's
case in their hands.

The judge paid Mr Armitage a warm compliment
on his speech. He agreed with much that he had
said. But, as regards his remarks on circumstantial
evidence, it was necessary for the jury to reflect that,
possibly, in fifteen or sixteen cases of murder out of
twenty, the only evidence which can be brought is
what is termed circumstantial; very seldom is a
human eye the actual witness of a homicide's crime.

He went through the evidence with sufficient im-
partiality, yet could not refrain from observing that,
while it seemed to him that much of it might have
been rebutted, if rebuttal had been possible, no attempt to do anything of the kind had been made.
Still it was true that if there was room for reasonable doubt as to the guilt of the man in front of them
it was their duty to acquit him. It was for them to
weigh the matter carefully in their own minds, in-

clining neither to the one side nor the other, and to say, without fear or favour, if there was such a doubt.

The jury retired to consider their verdict. The judge also withdrew. After an hour, returning into court, he sent a messenger to inquire if they were agreed. The answer came back that they were not. After an interval of another hour he sent again. The same reply was returned. It was then about seven o'clock. He directed the messenger to learn if there was any likelihood of their shortly agreeing, as, in that case, he would continue to sit, since it was of importance that the case should be concluded that night. There came a message to the effect that, if they were allowed a few minutes longer, they might arrive at an agreement.

Presently the twelve filed into court. The foreman announced that they agreed on their verdict. It was their unanimous opinion that the prisoner was guilty. The foreman's voice shook as he proclaimed the fact. He kept his eyes averted from the dock. He was a very young man, probably under thirty.

The prisoner was asked if he had anything to say why sentence of death should not be pronounced upon him. Someone at the back of the court exclaimed,—

"I have something to say."

All eyes were turned to discover the speaker. Even the foreman of the jury found courage to glance in the direction from which the voice had come.

CHAPTER XXIX

THE TRUE STORY OF WHAT TOOK PLACE IN EMBANK-MENT CHAMBERS

"Who was it that spoke?"

The question was asked by the judge. He looked angrily through his spectacles towards the spectators who were crowded together against the wall. A figure was threading its way towards the body of the court, the figure of a man. People made way for him to pass.

"I spoke, my lord."

It was Blaise Polhurston. He stood at counsel's table, looking across at the judge.

"Who are you, sir?"

"I am Blaise Polhurston. I am known to several persons who are present."

"Well, sir, is that any reason why you should interrupt the proceedings of the court?"

"You asked the prisoner if he had anything to say why sentence of death should not be pronounced upon him. I have something to say."

The judge addressed John Armitage.

"Do you know this person?"

John Armitage stood up.

"I do, my lord."

"If he has any evidence to give, why was he not called in the proper course?"

"I was not aware that he had any evidence to give."

John Armitage leant over towards Blaise Polhurston. That gentleman waved him back.

"With your permission, I would prefer to say what
I have to say to you direct, my lord, and to the jury,
without the intervention of any third person."

The judge regarded him with severity.

"Is what you have to say of capital importance?
I would warn you that your conduct is most irregular
and that it is a very serious matter to interrupt the
court at this stage of the proceedings."

"I am aware of it, my lord. What I have to say
is of the first importance."

"Then go into the box, and I will hear you. Let
this person be sworn."

The oath was duly administered. The judge himself conducted the examination, such as it was.

"Now, sir, what have you to say—briefly?"

Blaise Polhurston spoke clearly and quietly, without hesitation, straight on. Although there was still
that whimsical half-twinkle in his eyes, he looked
pale and worn, older than he had done of late. He
stooped as if he were tired. It was curious to notice
how all those who were present seemed to be domi-
nated by his quietly-spoken words.

"On the twelfth of December last I was penniless.
One whom I held dear was starving. On the afternoon of that day I had seen Howard Shapcott. Once
I had known him well, though it was years since we
had met."

"Had you parted on good terms?"

"We had not. Nor did we meet upon good terms.
He laughed at the plight which I was in, told me his
address, and jeeringly asked me to come and call on
him. When I returned to my lodging, I found that
the person to whom I have referred was, so far as I
could judge, dying for want of food. I had no money
with which to buy. For some days I had tried in

vain to get some. I resolved to ask a dole from Howard Shapcott. On my way to him, in the hope that I might still be spared that last humiliation, I called in at a common lodging-house, trusting that I might find there some acquaintance from whom I could obtain a few coppers. There was not a creature about the place. The kitchen was deserted. As I was going out, I saw that there was something on the mantelshelf. It was a revolver."

There was an exclamation from the prisoner in the dock,—

"That's how it was! I remember now that I left it on the shelf. That's what I done with it, and to think it should have come near to hanging me!"

The movement as of relief which passed through the court showed what a degree of tension had been broken by the prisoner's words. The judge was minatory.

"Silence, sir! How dare you interrupt like that?"

Yet one felt that his anger was an affair of the lips rather than of the heart. Blaise Polhurston went on.

"I put the revolver into my pocket. I can give no reason why I did so; it was done on the impulse of the moment; and I proceeded on my way to Howard Shapcott."

The judge interposed.

"I feel it my duty, at this stage, to warn you to be careful as to what statements you may make, as you are in no way bound to say anything which may tend to incriminate yourself."

"I thank you, my lord; I am aware of what are my legal obligations. And I will admit that I would not say what I am about to say had it not become necessary to do so, if an innocent man is to be saved from the gallows. When I arrived at Embankment Chambers I found no one in the entrance hall."

" What time was that ? "

" I should imagine about ten o'clock."

" The hall porter has sworn that he never left his post."

" I found no one there. Had I done so I take it that such a figure as I presented would have been questioned, if not refused admission. Shapcott had told me that his number was 212. I went straight up the stairs in search of it; to discover that I had set myself by no means an easy task. I wandered in all directions, up and down one corridor after another—"

" Without meeting anyone ? "

" Without meeting anyone. At last I found 212. Howard Shapcott's name was painted on the panel. I did not knock; I turned the handle, found the door was open, and walked in."

He paused for a moment. Some of those who were listening apparently took advantage of the opportunity to draw breath.

" Shapcott was seated at a writing-table. As I entered he rose to greet me. When he saw that it was I, moving quickly past me he locked the door."

" It was he, then, who locked the door on the inside."

" It was. He said as he did so, 'Since you have favoured me with this visit, we will secure ourselves from interruption.' I think that if I had known what he intended to do, I should have prevented him ; but his rapidity of movement took me aback."

Blaise Polhurston paused again, as if to arrange his thoughts.

" I find it difficult to narrate clearly what took place between us; to place events in their exact sequence. I have endeavoured often to do so to myself, but never to my satisfaction. I can only tell you the tale as I remember it. I saw at once that he was in a malicious mood — full of the old hatred for me, which feeling I returned in kind.

"But I was in extremity — cold, wet, hungry, footsore, penniless, in rags; and the memory of the plight of the one I had left at home was tearing at my heartstrings and made of me a coward. He was a prosperous gentleman, with the ball at his feet, at his ease, well fed—he told me he had been dining like a prince and drinking like a king. His tongue was like a rapier; mine never had been a match for his; then it was less so than ever. He ran me through and through; pricked me where he pleased; sought out my raws and galled them; while, so abject was my condition, that I was incapable of even attempting to return the blows he rained on me. The apathy with which I took my punishment did not provide him with sufficient entertainment. He made an effort to rouse me to resentment; so he began to make sport of a great wrong which he had done me in the days of my youth, and his, and for which, at the time, I could have killed him—and would have, if I had had the chance. The villain!"

For the first time the speaker showed a trace of passion. The epithet, though he still spoke quietly, was uttered with a sincerity which seemed to make it strike his hearers in the face.

"He threw his wickedness at me, by way of a gibe, like the scoundrel that he was; for he was a scoundrel, I say it though he is dead; if I meet him in the shades I will say it to his ghost. I felt something in the pocket of my coat. It was the revolver which I had taken from the mantelshelf. I took it out and showed it to him, and asked him how he dared to speak to me like that when, if I chose, at any moment I could shoot him like a dog. He laughed, exclaiming that he was quite easy in his mind, since that sort of thing

was not in my way, or I should have done it long
ago; which, in a sense, was true enough.

"I found that, after all, I could not ask for the
charity, the desperate hope of which had brought me
to him. I felt that, if I did ask, it would be refused.
But I could not ask. And since the jeering hideous
allusion to the evil, which I and others had suffered
at his hands, had served to stir even my sodden pulses,
I turned to leave the room.

"But he would not have it. From his point, perhaps, the sport had only just commenced. He saw
plenty more of it to follow. His appetite was
whetted, he would not have it baulked. As I turned
he sprang forward, caught me by the shoulder,
swung me round. That picture is ever before me.
I see him moving towards me, gripping my right
shoulder, swinging me round. The rest is dark.
That is all that I remember."

" What do you mean by saying that that is all you
remember? Though, mind, you are under no ob-
ligation to utter another word. What you do say
will be said voluntarily, after being duly warned
that, for you, the consequences may be most serious.
I want you to understand that quite distinctly."

"I do understand. I do not speak without consideration."

"Very good. So long as you properly appreciate
your responsibility for your own words. Now have
you anything further which you wish to say."

"I have. At this point in my mental consciousness
there seems to have been supervened an interregnum—
a hiatus, which I have endeavoured again and again
to fill. The next thing which I remember is that I
awoke out of what appeared to be a state of torpor, to
find myself lying on the floor of what seemed to be a
strange room. It was only after I had laboriously
gained my feet that I realised that the room was

Shapcott's. My brain was in confusion. I was giddy — objects were whirling round before my eyes; my limbs were trembling — I had to lean against the table to keep myself upright; I clearly recall how the sharp corner of the table penetrated the palm of my hand. Many moments must have elasped before I was able to even remotely appreciate the true inwardness of my position.

"As my vision, mental and physical, grew clearer, I perceived that, immediately in front of me, something was lying on the floor. I had been dimly wondering what had become of Shapcott. Here he was upon the floor. I wondered what he was doing there. 'Shapcott!' I said. 'Shapcott!' He did not answer. 'Shapcott!' I repeated, 'why don't you speak to me? What is the matter?'

"Again no reply. His continued silence began to impress even my blurred comprehension as ominous. Endeavouring to collect my thoughts, I regarded him with more attention, and was struck by the rigidity of his attitude, his complete quiescence, despite the uncomfortable posture in which he lay and the fashion in which his limbs were twisted. I bent over him. I touched him. He was dead.

"When I realised that this was so, I asked myself, in my surprise and my bewilderment, how the thing had come about. The only explanation which either then or since I have been able to supply I offer you. All my life I have been subject to an affection of the heart, which has caused me on very many occasions, generally at moments of unusual excitement, suddenly, without the slightest warning, to become completely unconscious. I have continued sometimes for hours together to be like one dead. I can only conjecture that, when Shapcott gripped me by the shoulder, I was seized by one of these attacks, that I fell, that, in falling, the revolver which I was holding in my

hand was fired, and that its discharge killed him on
the spot. Whether the responsibility for this was his
or mine I have been unable to finally decide; it
seems to me that it must have been purely acci-
dental.

"I think it possible that your lordship may consider that, the facts being as I have narrated them,
I ought to have gone straightway and proclaimed them
to the first person I might meet. That also may be
the opinion of the gentlemen of the jury. I did not
do so. In the first place, although consciously I had
had no hand in killing Howard Shapcott, I did not
regret that he was dead. Nor have I regretted it
since. In the second place, I had cut myself adrift
from my family; I had been parted from them for
years; I had sunk lower and lower in the social scale.
I did not propose, if I could help it, to allow the fact
of my continued existence to be brought to their
notice. In the third place, the story which I had to
tell was a strange one—as you will have perceived.
It quite probably would not have been credited. I
did not choose to subject myself to the personal in-
convenience to which, in any case, its telling would
have subjected me. And, in the last place, there was
someone lying at home who quite probably would die
if I did not hasten back with succour.

"Some coins were lying on the floor. They had
probably fallen from Shapcott's pocket. I do not know
what was the exact amount, but whatever it was, I
snatched them up. A ring and bracelet also lay close
by him. I took them too. Opening one of the French
windows, I saw that there was a balcony without.
Stepping on to it, I pulled the window to behind me.
Moving along the balcony, I found that at the window
of the next room the blind was up. Just inside, a
man with a black beard was standing, his face against
the pane. I felt that he must have seen me as I

passed. I was seized with a panic of terror. I rushed frantically on. At the end of the balcony was a waste water-pipe, fastened to the wall.

"By its aid I descended to the ground. I am no athlete. I have surveyed the scene since, and it has been strongly borne in upon me that the most mysterious part of the night's proceedings is how, by the employment of such means, I could have safely reached the ground.

"Nothing which I have said has been offered in extenuation of my conduct. I have no wish to excuse myself. As clearly and briefly as I could I have laid before you the facts of the matter as they are known to me. Up to now they have been known to me only. I should have continued to keep my own confidence had it not been for the verdict of the jury on Robert Foster. It has not been part of my intention to allow him to suffer for a crime which was no crime, and of which he, in any case, was wholly and completely innocent. That, my lord and gentlemen of the jury, is all I have to lay before you."

The silence which followed the speaker's words was broken by the sound of a feminine voice. A door was thrown open; a woman came hastening into the court.

"Gentleman! Gentleman!" she cried. "I heard you speaking! I knew it was you!"

Before anyone could stop her, she had pushed her way through the people to where Blaise Polhurston stood. It was Pollie Hills. He turned as she came and took her in his arms before them all.

"Pollie! 'We meet at Philippi!'"

Although it was unlikely that she gathered his exact import, it was plain that her guess came pretty near to it. Her tone was tremulous.

"What are you doing here? You haven't been telling them anything?"

"Yes, Pollie, I have. I've told them everything—the whole strange story."

"Gentleman! Gentleman!"

The usher's voice was heard.

"Silence! Order there!"

The judge leaned over the bench.

"What is the meaning of these extraordinary proceedings? Take that woman away!"

Pollie turned and screamed at him.

"You touch me! You let anyone so much as lay a hand on me! You let 'em dare!"

"Look out!" exclaimed a voice. "He's going to fall."

The allusion was to Blaise Polhurston, who did fall. Pollie caught him in her arms. She looked at him with a blend of amazement, horror, fear.

"Gentleman!" she gasped. "What's wrong?"

There was nothing wrong. He was only dead. That affection of the heart of which he had been telling them had visited him for the last time.

CHAPTER XXX

THE TESTAMENTARY DISPOSITIONS OF A GENTLEMAN OF FORTUNE

IT was found that Blaise Polhurston had left a will, which had been made on the day he had left Polhurston—the Saturday preceding that eventful Monday. The circumstances under which the document had come into being were clearly stated—in the expectation that Robert Foster would be found guilty, and that therefore he, the testator, would be constrained to take certain steps which might quite possibly result in his own demise. The steps had been taken, death had followed; expectation had been realised to the full. That he had foreseen that the end would come in the actual form it did was not certain, but it was probable; for it was shown that he had visited a physician who, perceiving that the shears were already nearly closed upon the thread of his life, had told him frankly that its severance was close at hand.

Five hundred pounds were left to Robert Foster, "whom I have unintentionally injured." A thousand were bequeathed to Pollie Hills, "the truest friend that ever a man had." A set of diamond ornaments to "my niece, Dolly Hamilton." These were found to have been purchased on the day on which the will was dated, and were discovered in an enclosure which was endorsed, " For my niece, Dolly, in fulfilment of a promise." The residue of all that he had went to Helen Fowler, "The daughter of the woman for whom I have esteemed my life well lost." The provision was safe-guarded to the best of his ability, Henry Baynes being named as guardian, and as sole executor. The will contained no mention of either his mother or his sister.

On the morning after the receipt of the news of her son's decease, Mrs Polhurston was found dead in bed. She, also, had died from heart's failure. Mrs Hamilton is, at present, in sole possession of the family estates. Her daughter, Dolly, is the wife of that "rising" young barrister, John Armitage, whose name, it is whispered, will be included in the next batch of "silks."

Pollie Hills is Mrs Robert Foster. In view of the way in which her false witness brought Robert Foster to the very foot of the gallows, the match seems a curious one ; but human nature is curious; seldom logical. It is certain that he forgave her even for her attempt against his life; it is possible that she was moved by his capacity for forgiveness even more than by his love ; it is sure that they joined forces, his five hundred to her thousand, and that now, as man and wife, they are joint proprietors of three or four flourishing establishments for the sale of fruit and vegetables. With them romance has been merged in actuality ; it is probable that already they are nearly oblivious of the fact that, for them, there ever was a period of "storm and stress."

The mists of romance still envelop Helen Fowler. After such a beginning, who can say, what is likely to be her end ? We may at least hope that she is destined for a happier fate than her mother. She is still young ; her beauty grows greater ; she has wealth ;

for such an one what radiant good fortune may the future not have in store!

And Blaise Polhurston is at peace. "After life's fitful fever, he sleeps well."

Rest assured.

THE END

Colston & Coy. Limited Printers, Edinburgh.

F. V. WHITE & Co.'s
Catalogue of Publications

ALEXANDER (Mrs)
Barbara, Lady's Maid and Peeress. 2/6.
Mrs Crichton's Creditor. 2/6 and 2/-. | A Fight with Fate. 2/6.
A Golden Autumn. 2/6 and 2/-.

ATLEE (H. FALCONER)
The Seasons of a Life. 6/-. | A Woman of Impulse. 6/-.

ALLEN (GRANT)
A Splendid Sin. 3/6.

ARNOLD (A. W.)
The Attack on the Farm. (Illustrated.) 6/-.

ARMSTRONG (Mrs)
Letters to a Bride. 2/6. | Good Form. 2/-, cloth.

BOUSTEAD (LEILA)
The Blue Diamonds. 1/- and 1/6.

CAMERON (Mrs LOVETT)
Devil's Apples. 2/6. | In a Grass Country. 1/- and 3/6.
A Man's Undoing. 2/6 and 2/-. | The Man Who Didn't. 1/- and 1/6.
Two Cousins and a Castle. 3/6.

CARRODER (CONRAD H.)
A Bride of God. 6/-.

CLEEVE (LUCAS)
The Monks of the Holy Tear. 6/-.

CROMMELIN (MAY)
Devil May Care. 6/-.

14 BEDFORD STREET, STRAND, W.C.

CROSS (MARY F.)
Railway Sketches. 1/-.

FARJEON (B. L.)
Miriam Rozella. 6/-.
The March of Fate. 2/6 and 2/-.
Basil and Annette. 2/6 and 2/-.

A Young Girl's Life. 2/6
Toilers of Babylon. 2/-.

FRASER (Mrs ALEXANDER)
A Modern Bridegroom. 2/-.

FETHERSTONHAUGH (The Honourable Mrs)
Dream Faces. 2/6.

GUNTER (ARCHIBALD CLAVERING)
A Florida Enchantment. 6/-.

GRIFFITH (GEORGE)
The Great Pirate Syndicate. 3/6.
The Destined Maid. 6/-.
Romance of Golden Star. (Illustrated.) 3/6 and 2/-.
Gambles with Destiny. 3/6.
The Gold Finder. (With Frontispiece.) 3/6 and 2/-.

GRAY (ANNABEL)
Forbidden Banns. 6/-.

GUBBINS (NATHANIEL)
Pink Papers. 1/-.

GORDON (LORD GRANVILLE)
The Race of To-day. 6/-. | Warned Off. 6/.

GRAEME ALASTOR (Mrs F. T. MARRYAT)
Romance of the Lady Arbell. 6/-.

GOWING (Mrs AYLMER)
Gods of Gold. 6/-. | Merely Players. 6/-.

HUMBERT (MABEL)

Continental Chit Chat. 1/-.

HUMPHRY (Mrs) ("MADGE," of *Truth*)

Housekeeping. 8/6.

JOCELYN (The Honourable Mrs)

Henry Massinger. 6/-.
Miss Rayburn's Diamonds. 6/-.
Lady Mary's Experiences. 2/6 & 2/-.

Only a Flirt. 2/6 and 2/-.
A Regular Fraud. 2/6 and 2/-.
Only a Horse Dealer. 2/6 and 2/-.

KENNARD (Mrs EDWARD)

At the Tail of the Hounds. 2/6.
A Riverside Romance. 2/6 & 2/-.
Fooled by a Woman. 2/6.
The Plaything of an Hour, and
other Stories. 2/6 and 2/-.
The Catch of the County. 2/6 and
2/-.
Wedded to Sport. 3/6 and 2/-.
The Hunting Girl. 2/6 and 2/-.
Just Like a Woman. 2/6 and 2/-.
Sporting Tales. 2/6 and 2/-.
Twilight Tales. (Illustrated.) 2/6.
That Pretty Little Horsebreaker.
2/6 and 2/-.

Matron or Maid? 2/-.
A Crack County. 2/6 and 2/-.
Landing a Prize. 2/6 and 2/-.
A Homburg Beauty. 2/6 and 2/-.
Our Friends in the Hunting Field.
2/6 and 2/-.
The Mystery of a Woman's Heart.
1/- and 1/6.
The Sorrows of a Golfer's Wife. 2/-.
A Guide Book for Lady Cyclists. 1/-
an 1/6.
The Girl in the Brown Habit. 3/6.
(Illustrated.)

LE QUEUX (WILLIAM)

England's Peril. 6/-.
The Veiled Man. 3/6.
Day of Temptation. 6/-.
Scribes and Pharisees. 3/6.
If Sinners Entice Thee. 6/-.
Zoraida. (Illustrated.) 3/6.

The Great War in England in 1897.
(Illustrated.) 3/6.
The Eye of Istar. (Illustrated.)
3/6 and 2/-.
Whoso Findeth a Wife. 3/6.
Devil's Dice. 3/6.

LYALL (J. G.)

Norrington Le Vale. 6/-.

The Merry Gee-Gee. 2/6.

MITFORD (BERTRAM)
The Weird of Deadly Hollow. 3/6.
The Ruby Sword. (Illustrated.) 3/6.

MEADE (Mrs L. T.)
The Siren. 6/-.
A Son of Ishmael. 2/-, cloth.

The Way of a Woman. 2/6 and 2/-.

MIDDLEMASS (JEAN)
Blanche Coningham's Surrender. 6/-.

MARSH (RICHARD)
The House of Mystery. 6/-.

MELL (F. H.)
The Gods Saw Otherwise. 6/-.

McMILLAN (Mrs ALEC)
The Evolution of Daphne. 6/-.

MATHERS (HELEN)
A Man of To-day. 2/6 and 2/-.

MARRYAT (FLORENCE)
A Rational Marriage. 6/-.

The Spirit World. 2/-.

NISBET (HUME)
For Liberty. 3/6 and 2/-.
A Sweet Sinner. 3/6 and 2/-.
The Rebel Chief. 3/6 and 2/-.
A Desert Bride. 3/6 and 2/-.
A Bushgirl's Romance. 3/6 & 2/-.

Comrades of the Black Cross. 3/6.
The Queen's Desire. 3/6 and 2/-.
The Great Secret. 2/6 and 2/-.
The Haunted Station. 2/6.
My Love Noel. 2/6 and 2/-.

PRAED (Mrs CAMPBELL)
The Romance of a Chalet. 2/6 and 2/-.

PHILIPS (F. C.) and C. J. WILLS
Sybil Ross's Marriage. 2/6 and 2/-.

PRIMM (PERRINGTON)
Belling the Cat. 6/-.

"RITA."

Vignettes. 2/6 and 2/-.
Joan & Mrs Carr. 2/6 and 2/-.

The Man in Possession. 2/6 & 2/-.
Miss Kate. 2/6.

RIDDLE (Mrs)

Handsome Phil. 3/6.

A Rich Man's Daughter. 2/6 and 2/-.

ROGERS (C. V.)

Her Marriage Vow. 6/-.

SIMS (G. R.)

As it Was in the Beginning. 2/6 and 2/-.
The Coachman's Club. 2/6 and 2/-.
Dorcas Dene, Detective. 1/- and 1/6, cloth.
Dorcas Dene, Detective. (Second Series.) 1/- and 1/6.

ST AUBYN (ALAN)

A Proctor's Wooing. 6/- and 2/-, cloth.
Bonnie Maggie Lauder. 6/-.
To Step Aside is Human. 2/6 and 2/-.

The Wooing of May. 3/6 and 2/-.
A Tragic Honeymoon. 2/.
In the Sweet West Country. 2/6 and 2/.

STUART (ESMÈ)

Arrested. 6/-.

The Strength of Two. 6/-.

SERGEANT (ADELINE)

The Love Story of Margaret Wynne. 6/-.
A Valuable Life. 6/-.

Told in the Twilight. 2/6 and 2/-.
In Vallombrosa. 3/6.

SAVAGE (RICHARD HENRY)

An Awkward Meeting. 2/6 and 2/-.

THOMAS (ANNIE)

Essentially Human. 6/-.

WARDEN (FLORENCE)

The Secret of Lynndale. 6/-.
The Bohemian Girls. 6/-.
Girls will be Girls. 6/-.
Our Widow. 2/6 and 2/-.
A Lady in Black. 2/6 and 2/-.

A Spoilt Girl. 2/-.
The Girls at the Grange. 2/6.
The Mystery of Dudley Horne. 2/6 and 2/-.

WINTER (JOHN STRANGE)

A Summer Jaunt. 3/6.

Heart and Sword. 6/-.

The Price of a Wife. 3/6.

The Peacemakers. 2/6.

Everybody's Favourite. 3/6 and 2/-.

Into an Unknown World. 2/6 and 2/-.

The Strange Story of My Life. 2/6 and 2/-.

Truth Tellers. 2/6 and 2/-.

A Magnificent Young Man. 2/6 and 2/-.

A Blameless Woman. 2/6 and 2/-.

A Born Soldier. 2/6 and 2/-.

A Seventh Child. 2/6 and 2/-.

The Soul of the Bishop. 2/6 and 2/-.

Aunt Johnnie. 2/6 and 2/-.

My Geoff. 2/6 and 2/-.

The Binks Family. 1/- and 1/6.

Sentimental Maria. 1/- and 1/6·

The Same Thing with a Difference. 1/- and 1/6.

I Married a Wife. (Profusely Illustrated.) 1/- and 1/6.

Private Tinker; and other Stories. (Profusely Illustrated.) 1/- and 1/6.

The Major's Favourite. 1/- and 1/6.

The Stranger Woman. 1/- and 1/6.

Red Coats. (Profusely Illustrated.) 1/- and 1/6.

A Man's Man. 1/- and 1/6.

That Mrs Smith. 1/- and 1/6.

Three Girls. 1/- and 1/6.

Mere Luck. 1/- and 1/6.

Lumley the Painter. 1/- and 1/6.

Good-Bye. 1/- and 1/6.

Only Human. 2/6 and 2/-.

The Other Man's Wife. 2/6 and 2/-.

Army Society. 2/-.

A Siege Baby. 2/6 and 2/-.

Beautiful Jim. 2/6 and 2/-.

Garrison Gossip. 2/6 and 2/-.

Mrs Bob. 2/6 and 2/-.

A Gay Little Woman. 1/- and 1/6.

A Seaside Flirt. 1/- and 1/6.

The Troubles of an Unlucky Boy. 1/- and 1/6.

Grip. 1/- and 1/6.

I Loved Her Once. 1/- and 1/6.

He Went for a Soldier. 1/- and 1/6.

Ferrers Court. 1/- and 1/6.

A Little Fool. 1/- and 1/6.

Buttons. 1/- and 1/6.

Bootles' Children. (Illustrated.) 1/- and 1/6.

The Confessions of a Publisher. 1/- and 1/6.

My Poor Dick. (Illustrated.) 1/- and 1/6.

That Imp. 1/- and 1/6.

Mignon's Secret. 1/- and 1/6.

Mignon's Husband. 1/- and 1/6.

On March. 1/- and 1/6.

In Quarters. 1/- and 1/6.

In the Same Regiment. 1/- & 1/6.

Two Husbands. 1/- and 1/6.

YOLLAND (E.)

In Days of Strife. 6/-. | Mistress Bridget. 6s.

SIX SHILLING NOVELS

In One Vol., Cloth Gilt, Price 6s. each

ENGLAND'S PERIL. By William Le Queux.

HEART AND SWORD. By John Strange Winter.

THE DAY OF TEMPTATION. By William Le Queux.

A VALUABLE LIFE. By Adeline Sergeant.

A RATIONAL MARRIAGE. By Florence Marryat.

THE DESTINED MAID. By George Griffith.

IF SINNERS ENTICE THEE. By William Le Queux.

HER MARRIAGE VOW. By C. V. Rogers.

THE BOHEMIAN GIRLS. By Florence Warden.

THE SECRET OF LYNNDALE. By the same Author.

DIVIL MAY CARE. By May Crommelin.

MISS RAYBURN'S DIAMONDS. By the Hon. Mrs Jocelyn.

HENRY MASSINGER. By the same Author.

THE MONKS OF THE HOLY TEAR. By Lucas Cleeve. Author of "Lazarus," &c.

WARNED OFF. By Lord Granville Gordon.

THE SIREN. By L. T. Meade.

THE HOUSE OF MYSTERY. By Richard Marsh, Author of "The Beetle."

THE SEASONS OF A LIFE. By H. Falconer Atlee.

A WOMAN OF IMPULSE. By the same Author.

BLANCHE CONINGHAM'S SURRENDER. By Jean Middlemass.

SIX SHILLING NOVELS—Continued

THE STRENGTH OF TWO. By Esmè Stuart.

MIRIAM ROZELLA. By B. L. Farjeon.

LADY MARY'S EXPERIENCES. By Mrs Robert Jocelyn.

MERELY PLAYERS. By Mrs Aylmer Gowing.

THE RACE OF TO-DAY. By Lord Granville Gordon.

BONNIE MAGGIE LAUDER. By Alan St Aubyn.

GIRLS WILL BE GIRLS. By Florence Warden.

THE EVOLUTION OF DAPHNE. By Mrs Alec McMillan.

GODS OF GOLD. By Mrs Aylmer Gowing.

ARRESTED. By Esmè Stuart.

THROUGH THE BUFFER STATE: A Record of Recent Travels through Borneo, Siam and Cambodia. By Surgeon-Major Macgregor, M.D. Ten whole-page Illustrations and a Map.

MISTRESS BRIDGET. By E. Yolland.

LITTLE MISS PRIM. By Florence Warden.

THE LOVE STORY OF MARGARET WYNNE. By Adeline Sergeant.

THE ATTACK ON THE FARM. By A. W. Arnold. (Illustrated.

A BRIDE OF GOD. By Conrad H. Carroder.

FORBIDDEN BANNS. By Annabel Gray.

NORRINGTON LE VALE. By J. G. Lyall.

ROMANCE OF THE LADY ARBELL. By Alastor Graeme (Mrs F. T. Marryat).

BELLING THE CAT. By Perbington Primm.

A FLORIDA ENCHANTMENT. By Archibald Clavering Gunter.

THE GODS SAW OTHERWISE. By F. H. Mell.